# Editorial Policy

§ 1. Lecture Notes aim to report new developments - quickly, informally, and at a high level. The texts should be reasonably self-contained and rounded off. Thus they may, and often will, present not only results of the author but also related work by other people. Furthermore, the manuscripts should provide sufficient motivation, examples and applications. This clearly distinguishes Lecture Notes manuscripts from journal articles which normally are very concise. Articles intended for a journal but too long to be accepted by most journals, usually do not have this "lecture notes" character. For similar reasons it is unusual for Ph. D. theses to be accepted for the Lecture Notes series.

§ 2. Manuscripts or plans for Lecture Notes volumes should be submitted (preferably in duplicate) either to one of the series editors or to Springer- Verlag, Heidelberg . These proposals are then refereed. A final decision concerning publication can only be made on the basis of the complete manuscript, but a preliminary decision can often be based on partial information: a fairly detailed outline describing the planned contents of each chapter, and an indication of the estimated length, a bibliography, and one or two sample chapters - or a first draft of the manuscript. The editors will try to make the preliminary decision as definite as they can on the basis of the available information.

§ 3. Final manuscripts should preferably be in English. They should contain at least 100 pages of scientific text and should include

- a table of contents;
- an informative introduction, perhaps with some historical remarks: it should be accessible to a reader not particularly familiar with the topic treated;
- a subject index: as a rule this is genuinely helpful for the reader.

Further remarks and relevant addresses at the back of this book.

# Lecture Notes in Mathematics 1694

Springer
*Berlin*
*Heidelberg*
*New York*
*Barcelona*
*Budapest*
*Hong Kong*
*London*
*Milan*
*Paris*
*Singapore*
*Tokyo*

Andrea Braides

# Approximation of Free-Discontinuity Problems

Springer

Author

Andrea Braides
SISSA (International School for Advanced Studies)
via Beirut 4
I-34013 Trieste, Italy
e-mail: braides@sissa.it

Cataloging-in-Publication Data applied for

**Die Deutsche Bibliothek - CIP-Einheitsaufnahme**

**Braides, Andrea:**
**Approximation of free discontinuity problems / Andrea Braides. - Berlin ; Heidelberg ; New York ; Barcelona ; Budapest ; Hong Kong ; London ; Milan ; Paris ; Santa Clara ; Singapore ; Tokyo : Springer, 1998**
**(Lecture notes in mathematics ; 1694)**
**ISBN 3-540-64771-6**

Mathematics Subject Classification (1991): 49J45, 35B20, 73M25

ISSN 0075-8434
ISBN 3-540-64771-6 Springer-Verlag Berlin Heidelberg New York



Typesetting: Camera-ready TEX output by the author
SPIN: 10650077 41/3143-543210 - Printed on acid-free paper

In memory of my mother

# PREFACE

In recent years much attention has been devoted to variational problems involving bulk and surface energies at the same time, with applications to the study of multi-phase systems, Fracture Mechanics, Computer Vision, etc. A weak formulation of some of these problems has been proposed by De Giorgi and Ambrosio by introducing the space of special functions of bounded variation. In their approach, the unknown surface of the problem is interpreted as the discontinuity set of an unknown function (from which the terminology "free-discontinuity problems"). Their theory has led to many existence and regularity results; at the same time, many approximations have been proposed to obtain smooth approximate solutions of these problems to overcome the difficulties arising from the presence of an unknown discontinuity surface. Purpose of these lecture notes is to present an unifying view of all these approximation procedures in the framework of $\Gamma$-convergence.

In Chapters 1 and 2 we briefly present a self-contained introduction to functions of bounded variation and to the existence theory in SBV spaces. In Chapter 3 we give the necessary definitions about $\Gamma$-convergence and present the 1-dimensional version of all the approximation procedures. Chapter 4 is devoted to a general approach to approximations by slicing and density techniques, which allows to reduce $n$-dimensional problems to the 1-dimensional ones studied in the previous chapter. Non-local approximations are dealt with separately in Chapter 5. Finally, some further issues connected to approximation problems are collected in the Appendix.

The content of these notes has formed the core of an advanced course given by the author at SISSA, Trieste in 1996/97, directed to Ph. D. students in Functional Analysis and Applications. The author is grateful to all the students of the course who provided a lively and stimulating interaction, and acknowledges the contributions of R. Alicandro, G. Cortesani, G. Dal Maso, A. Garroni, M.S. Gelli, M. Gobbino and A. Malchiodi to the content of these notes.

Part of the book was conceived and written during a visit of the author to the Max Planck Institute for Mathematics in the Sciences in Leipzig on a Marie Curie fellowship of the EU program "Training and Mobility of Researchers". This work was supported by *Consiglio Nazionale delle Ricerche* through the project *Equazioni alle derivate parziali e calcolo delle variazioni*.

Trieste, May 1998

# PREFACE

CONTENTS

# INTRODUCTION

Following a terminology introduced by De Giorgi, we denote as "free-discontinuity problems" all those problems in the calculus of variations where the unknown is a pair $(u, K)$ with $K$ varying in a class of (sufficiently smooth) closed hypersurfaces contained in a fixed open set $\Omega \subset \mathbf{R}^n$ and $u : \Omega \setminus K \to \mathbf{R}^m$ belonging to a class of (sufficiently smooth) functions. Such problems are usually of the form

$$\min\{E_v(u,K) + E_s(u,K) + \text{ "lower order terms"}\}, \tag{1}$$

with $E_v$, $E_s$ being interpreted as *volume* and *surface* energies, respectively. Several examples can be described in this setting. We list a few ones.

(i) *Signal reconstruction.* A source signal is usually a piecewise smooth function $u$ (which we may think as parameterized on some interval $(a, b)$) (see [St], [Mu]). The problem of reconstructing $u$ from a disturbed input $g$ deriving from a distorted transmission, can be modelled as finding the minimum

$$\min\Big\{\int_{(a,b)\setminus S(u)} |u'|^2\,dt + c_1 \int_a^b |u-g|^2\,dt + c_2\,\#(S(u))\Big\}, \tag{2}$$

where $S(u)$ denotes the set of discontinuity points of $u$. Here, $c_1$ and $c_2$ are tuning parameters. In this case $\Omega = (a, b)$, $K = S(u)$, $E_v(u, K) = \int_{\Omega\setminus K} |u'|^2\,dt$, and $E_s(u, K) = c_2\,\#(K)$.

(ii) *Image reconstruction.* The formulation above can be extended to model some problems in computer vision by introducing the functional (see [MS], [MoS], [DG], [DMS], [Sh])

$$\int_{\Omega\setminus K} |\nabla u|^2\,dx + c_1 \int_{\Omega\setminus K} |u-g|^2\,dx + c_2\mathcal{H}^1(K)\,. \tag{3}$$

In this case $g$ is interpreted as the input picture taken from a camera, $u$ is the "cleaned" image, and $K$ is the relevant contour of the objects in the picture. Again, $c_1$ and $c_2$ are contrast parameters. Note that the problem is meaningful also adding the constraint $\nabla u = 0$ outside $K$, in which case we have a minimal partitioning problem (see e.g. [MS], [CTa], [AB1], [AB2], [BCP]).

(iii) *Fractured hyperelastic media.* In this case $\Omega \subset \mathbf{R}^3$ is the reference configuration of an elastic body, $K$ is the crack surface, and $u$ represents the elastic deformation in the unfractured part of the body. Following Griffith's theory of fracture, we can introduce a surface energy which accounts for fracture initiation (see [Gr], [Ba], [AB3]). In the homogeneous case, this energy is simply proportional to the surface area of $K$ if the body is isotropic: $E_s(u, K) = c\mathcal{H}^2(K)$, or, more generally, $E_s$ is an integral on $K$ depending on the orientation $\nu$ of the crack surface in the non-isotropic case: $E_s(u,K) = \int_K \varphi(\nu)\,d\mathcal{H}^2$. The volume energy takes the form $E_v(u,K) = \int_{\Omega\setminus K} W(\nabla u)\,dx$, where $W$ is an elastic bulk energy density. (For free-discontinuity problems in this framework see, for example, [AB3], [BC], [BDV], [FF].)

(iv) *Drops of liquid crystals* (Oseen-Frank energy with surface interaction). In this model $D \subset \Omega$ represents the region occupied by a liquid crystal, whose energy is

$$\int_D W(n, \nabla n)\,dx + \int_{\partial D\cap\Omega} f(n, \nu_D)\,d\mathcal{H}^2 \,. \tag{4}$$

In this case, $K = \partial D$, $n$ is the orientation of the crystal, $u = (n, \chi_D)$, and $\nu_D$ is the normal to $\partial D$ (see [Vi]).

(v) *Prescribed curvature problems.* As a particular free-discontinuity problem we can also recover the problem of finding sets $E$ with boundary of prescribed mean curvature $H$; i.e., satisfying $H = g\nu$ on $\partial E$. In this case the energy to be minimized is

$$\int_E g(x)\,dx + \mathcal{H}^{n-1}(\partial E)\,, \tag{5}$$

which can be seen as the sum of a volume and a surface energy depending on the unknowns $u = \chi_E$ and $K = \partial E$.

The treatment of free-discontinuity problems following the direct methods of the calculus of variations presents many difficulties, due to the dependence of the energies on the surface $K$. Unless topological constraints are added, it is usually not possible to deduce compactness properties from the only information that such kind of energies are bounded. An idea of De Giorgi has been to interpret $K$ as the set of discontinuity points of the function $u$, and to set the problems in a space of discontinuous functions. The requirements on such a space are of two kinds:

(a) *structure properties*: if we define $K$ as the set of discontinuity points of the function $u$ then $K$ can be interpreted as an hypersurface, and $u$ is "differentiable" on $\Omega \setminus K$ so that bulk energy depending on $\nabla u$ can be defined;

(b) *compactness properties*: it is possible to apply the direct method of the calculus of variations, obtaining compactness of sequences of functions with bounded energy.

The answer to the two requirements above has been De Giorgi and Ambrosio's space of *special functions of bounded variation* [DGA]: a function $u$ belongs to $SBV(\Omega)$ if and only if its distributional derivative $Du$ is a bounded measure that can be split into a bulk and a surface term. This definition can be further specified: if $u \in SBV(\Omega)$ and $S(u)$ stands for the complement of the set of the Lebesgue points for $u$ then a measure-theoretical normal $\nu_u$ to $S(u)$ can be defined $\mathcal{H}^{n-1}$-a.e. on $S(u)$, together with the traces $u^\pm$ on both sides of $S(u)$; moreover, the approximate gradient $\nabla u$ exists a.e. on $\Omega$, and we have

$$(6) \qquad Du = \nabla u\, \mathcal{L}_n + (u^+ - u^-)\nu_u\, \mathcal{H}^{n-1} \llcorner S(u)\,.$$

Replacing the set $K$ by $S(u)$ we obtain a weak formulation for free-discontinuity problems, whose energies take the general form

$$(7) \qquad \int_\Omega f(x,u,\nabla u)\,dx + \int_{S(u)} \vartheta(x,u^+,u^-,\nu_u)\,d\mathcal{H}^{n-1}\,.$$

An important and extensively studied model case is given by the Mumford-Shah functional

$$(8) \qquad E(u) = \alpha \int_\Omega |\nabla u|^2\,dx + \beta\mathcal{H}^{n-1}(S(u))\,.$$

An existence theory for problems involving these kinds of energies has been developed by Ambrosio [A1]–[A4]. Various regularity results show that for a wide class of problems the weak solution $u$ in $SBV(\Omega)$ provides a solution to the corresponding free-discontinuity problem, taking $K = \overline{S(u)}$ (see [DGC], [CL], [AFP], [Bo], [Di]).

Despite the existence theory developed in SBV-spaces, functionals arising in free-discontinuity problems present some serious drawbacks. First, the lack of differentiability in any reasonable norm implies the impossibility of flowing these functionals, and dynamic problems can be tackled only in an indirect way. Moreover, numerical problems arise in the detection of the unknown discontinuity surface. To bypass these difficulties, a considerable effort has been spent recently to provide variational approximations of free discontinuity problems, and in particular of the Mumford-Shah functional $E$ defined in (8), with differentiable energies defined on smooth functions.

The natural notion of convergence for these types of problems has turned out to be that of De Giorgi's Γ-*convergence* (see [DGF], [DM], [BDF]). We recall that a family $(F_\varepsilon)$ of real-valued functions defined on a metric space $X$ Γ-converges to $F$ as $\varepsilon \to 0^+$ if for all $x \in X$

(i) (liminf inequality) $F(x) \le \liminf_{\varepsilon\to 0^+} F_\varepsilon(x_\varepsilon)$ if $x_\varepsilon \to x$;

(ii) (existence of a recovery sequence) $F(x) = \lim_{\varepsilon\to 0^+} F_\varepsilon(x_\varepsilon)$ for some $x_\varepsilon$ with $x_\varepsilon \to x$.

This notion enjoys useful compactness properties, and, under suitable equi-coerciveness assumptions, is strong enough to guarantee that minima and minimizers for problems related to $F_j$ converge to the corresponding minima and minimizers for problems related to $F$. The proof of the "liminf inequality" is usually the most technical part in a Γ-convergence result, while the form of "recovery sequences" gives an insight of the nature of the convergence.

A first approximation by Γ-convergence of the Mumford-Shah was given by Ambrosio and Tortorelli in [AT1] and [AT2]. Following an earlier idea developed by Modica and Mortola [MM], who approximated the perimeter functional by elliptic functionals, Ambrosio and Tortorelli introduced an approximation procedure of $E(u)$ with an auxiliary variable $v$, which in the limit approaches $1-\chi_{S(u)}$. A family of functional studied in [AT2] is the following:

$$G_\varepsilon(u,v) = \int_\Omega v^2|\nabla u|^2\,dx + \frac{1}{2}\int_\Omega \Big(\varepsilon|\nabla v|^2 + \frac{1}{\varepsilon}(1-v)^2\Big)\,dx\,, \tag{9}$$

defined on functions $u, v$ such that $v \in H^1(\Omega)$, $uv \in H^1(\Omega)$ and $0 \le v \le 1$, which $\Gamma$-converges as $\varepsilon \to 0+$ with respect to the $(L^1(\Omega))^2$-topology to the functional

$$G(u,v) = \begin{cases} E(u) & \text{if } v = 1 \text{ a.e. on } \Omega \\ +\infty & \text{otherwise,} \end{cases} \tag{10}$$

defined on $(L^1(\Omega))^2$. Clearly, the functional $G$ is equivalent to $E$ as far as minimum problems are concerned. As the functionals $G_\varepsilon$ are elliptic, even though non-convex, numerical methods can be applied to them (see [BeC], [Sh]). It is clear, though, that the introduction of an extra variable $v$ can be very demanding from a numerical viewpoint. Functionals of the form (9) can be modified, even though not in a straightforward way, to obtain more general energies (see [ABS], [AB]).

A simpler approach is to try an approximation by means of local integral functionals of the form

$$\int_\Omega f_\varepsilon(\nabla u(x))\,dx\,, \tag{11}$$

defined in the Sobolev space $H^1(\Omega)$. It is clear that such functionals cannot provide any variational approximation for $E$. In fact, if an approximation existed by functionals of this form, the functional $E(u)$ would also be the $\Gamma$-limit of their lower semicontinuous envelopes; i.e., the convex functionals

$$\int_\Omega f_\varepsilon^{**}(\nabla u(x))\,dx\,, \tag{12}$$

where $f_\varepsilon^{**}$ is the convex envelope of $f_\varepsilon$ (see, e.g., [DM], Proposition 6.1 and Example 3.11), in contrast with the lack of convexity of $E(u)$. However, functionals of the form (11) can be a useful starting point. We can begin by requiring that for every $u \in SBV(\Omega)$ with $\nabla u$ and $S(u)$ sufficiently smooth we have

$$\lim_{\varepsilon\to 0+}\int_\Omega f_\varepsilon(\nabla u_\varepsilon(x))\,dx = \alpha\int_\Omega |\nabla u|^2\,dx + \beta\mathcal{H}^{n-1}(S(u))$$

if we choose $u_\varepsilon$ to be very close to $u$, except in an $\varepsilon$-neighbourhood of $S(u)$ (where the gradient of $u_\varepsilon$ tends to be very large). It can be easily seen that this requirement is fulfilled if we choose $f_\varepsilon$ of the form

$$f_\varepsilon(\xi) = \frac{1}{\varepsilon}f(\varepsilon|\xi|^2), \text{ with } f'(0) = \alpha \text{ and } \lim_{t\to+\infty} f(t) = \frac{\beta}{2}\,. \tag{13}$$

Non-convex integrands of this form can be exploited, provided we slightly modify the functionals in (11). This can be done in many ways. For example, dealing for simplicity with the 1-dimensional case, the convexity constraint in $\nabla u$ can be removed by introducing a second-order singular perturbation, of the form

$$E_\varepsilon(u) = \frac{1}{\varepsilon}\int_\Omega f(\varepsilon|u'|^2)\,dx + \varepsilon^3\int_\Omega |u''|^2\,dx \tag{14}$$

on $H^2(\Omega)$. Note that the $\Gamma$-limit of these functionals would be trivial without the last term, and that the convexity in $u''$ assures the weak lower semicontinuity of $E_\varepsilon$ in $H^2(\Omega)$. In [ABGe] it has been proven that the family $(E_\varepsilon)$ $\Gamma$-converges to the functional defined on $SBV(\Omega)$ by

$$F(u) = \alpha\int_\Omega |u'|^2\,dx + C\sum_{t\in S(u)}\sqrt{|u^+(t)-u^-(t)|}\,, \tag{15}$$

with $C$ explicitly computable from $\beta$. This method can be generalized to higher dimension (see [AGe]). In contrast to those in (9), the functionals in (14) possess a particularly simple form, with no extra variable. The form of the approximating functionals gets more complex if we want to use this approach to recover in the limit other surface energies (as, for example, in the Mumford-Shah functional) in which case we must substitute $f$ by more complex $f_\varepsilon$ in (14).

A different path can be followed, considering approximations of the form

$$E_\varepsilon(u) = \frac{1}{\varepsilon}\int_\Omega f\Big(\varepsilon \fint_{B_\varepsilon(x)\cap\Omega} |\nabla u(y)|^2\,dy\Big)\,dx\,, \tag{16}$$

defined for $u\in H^1(\Omega)$, where $f$ is a suitable non-decreasing continuous (non-convex) function. These functionals are non-local in the sense that their energy density at a point $x\in\Omega$ depends on the behaviour of $u$ in the whole set $B_\varepsilon(x)\cap\Omega$. Note that, even if the term containing the gradient is not convex, the functional $E_\varepsilon$ is weakly lower semicontinuous in $H^1(\Omega)$ by Fatou's Lemma. These functionals $\Gamma$-converge, as $\varepsilon\to 0$ to the Mumford-Shah functional $E$ in (8) if $f$ satisfies the limit conditions in (13) (see [BDM]). Furthermore, by taking suitable $f_\varepsilon$ in place of $f$ in (16), one can approximate functionals with more general surface energy densities (see [BG]). Note, however, that by the nature of the approximating functionals, the limit surface energy density will always be an increasing function of the jump width.

A recent conjecture by De Giorgi, proved by Gobbino [Go], provides another type of non-local approximation of the Mumford-Shah functional (in the form (8) with suitable $\alpha,\beta$), with approximating functionals the family

$$E_\varepsilon(u) = \frac{1}{\varepsilon^{n+1}}\int_{\Omega\times\Omega}\arctan\Big(\frac{(u(x)-u(y))^2}{\varepsilon}\Big)\,e^{-|x-y|^2/\varepsilon}\,dx\,dy\,, \tag{17}$$

defined on $L^1(\Omega)$. This procedure is particularly flexible, allowing for easy generalizations, to approximate general functionals. The main drawback of this approach is the difficulty in obtaining coerciveness properties.

In the next chapters we provide an unifying view of all the approximation procedures outlined above in the framework of $SBV$ functions and $\Gamma$-convergence. The first chapter is devoted to the main results about $BV$-functions, of whom we give a brief self-contained account, referring to Ambrosio, Fusco and Pallara [AFP1], Evans and Gariepy [EG] or Federer [Fe] for a deeper insight. Chapter 2 is devoted to the theory of $SBV$ and $GSBV$ functions; here we prove the fundamental compactness and lower semicontinuity theorems. In Chapter 3 we begin to deal with approximation problems, introducing the notion of $\Gamma$-convergence of families of functionals, and proving all the approximation results in the 1-dimensional case. This central part is the most important one, as the $n$-dimensional proofs will often reduce to a 1-dimensional study. Moreover, the construction of recovery sequences in the 1-dimensional case gives a clarifying illustration of the approximation procedures. The passage to higher dimension is obtained in Chapter 4, where the most technical results are proved. It relies on two different methods to obtain upper and lower estimates: the lower inequality is derived by using a "slicing" technique, which allows to reduce to the 1-dimensional case, while the upper inequality is obtained by first constructing optimal sequences for special classes of functions and then proving a density result. Finally, Chapter 5 is fully devoted to the non-local approximations. We recover the proof of all the results mentioned in this Introduction; note, moreover, that many proofs are different from those available in the literature, and many results are new.

# 1

# FUNCTIONS OF BOUNDED VARIATION

This chapter provides the necessary background about measure theory and functions of bounded variation. Only the proofs relevant to subsequent reasonings are included, while hints to the proofs of most results are given in the exercises section at the end of the chapter. We refer to the books of Evans and Gariepy [EG], Ambrosio, Fusco and Pallara [AFP1], Giusti [Gi] and Federer [Fe] for a deeper insight in the subject.

## 1.1 Measure theory. Basic notation

In this section we give the main definitions of measure theory. We consider measures as set functions.

**Definition 1.1** *A function $\mu : \mathcal{B}(\Omega) \to \mathbf{R}^N$ is a (vector)* measure *on $\Omega$ if it is countably additive; i.e.,*

$$B = \bigcup_{i\in\mathbf{N}} B_i, \ B_i \cap B_j = \emptyset \text{ if } i \neq j \qquad \Longrightarrow \qquad \mu(B) = \sum_{i\in\mathbf{N}} \mu(B_i).$$

*The set of such measures will be denoted by $\mathcal{M}(\Omega;\mathbf{R}^N)$. If no confusion may arise, we denote by $\mu_i$ ($i = 1,\dots,N$) the components of $\mu$; i.e., we may write $\mu(B) = (\mu_1(B),\dots,\mu_N(B))$.*

*We say that a measure is a* scalar measure *if $N = 1$, and that it is a* positive measure *if it takes its values in $[0,+\infty)$. The sets of scalar and of positive measures will be denoted by $\mathcal{M}(\Omega)$ and $\mathcal{M}_+(\Omega)$, respectively.*

*A function $\mu : \mathcal{B}_c(\Omega) \to \mathbf{R}^N$ is a* Radon measure *on $\Omega$ if $\mu_{|\mathcal{B}(\Omega')}$ is a measure on $\Omega'$ for all $\Omega' \subset\subset \Omega$. As above, we will speak of scalar and of positive Radon measures.*

*If $\mu \in \mathcal{M}(\Omega)$ we adopt the usual notation $L^p(\Omega,\mu;\mathbf{R}^N)$ to indicate the space of $\mathbf{R}^N$-valued p-summable functions with respect to $|\mu|$ on $\Omega$. We omit $\mu$ if it is the Lebesgue measure, and we omit $\mathbf{R}^N$ if $N = 1$.*

**Remark 1.2** If $\mu$ is a positive Radon measure then we define

$$\mu(B) = \lim_h \mu(B \cap \Omega_h) \in [0,+\infty]$$

for all $B \in \mathcal{B}(\Omega)$, where $\Omega_h \subset\subset \Omega$ converges increasingly to $\Omega$.

**Definition 1.3** *Let $\mu : \mathcal{P}(\Omega) \to \mathbf{R}^N$ be a set function. We define then the* restriction $\mu \llcorner B$ *of $\mu$ to $B \subset \Omega$ by*

$$\mu \llcorner B(A) = \mu(B \cap A)$$

*for all $A \in \mathcal{P}(\Omega)$. We use the same notation if $\mu \in \mathcal{M}(\Omega; \mathbf{R}^N)$, in which case also $\mu \llcorner B(A)$ is defined on $\mathcal{B}(\Omega)$ and $\mu \llcorner B \in \mathcal{M}(\Omega; \mathbf{R}^N)$.*

**Remark 1.4** The Lebesgue measures on $\mathbf{R}^n$ can be defined as the unique positive Radon measure $\mathcal{L}_n$ on $\mathbf{R}^n$ satisfying $\mathcal{L}_n([0,1]^n) = 1$ and $\mathcal{L}_n(a + tA) = t^n \mathcal{L}_n(A)$ for all $a \in \mathbf{R}^n$, $A \in \mathcal{B}(\mathbf{R}^n)$, and $t > 0$. We also use the notation $|A| = \mathcal{L}_n(A)$.

**Definition 1.5** *If $\mu \in \mathcal{M}(\Omega; \mathbf{R}^N)$ for all $B \in \mathcal{B}(\Omega)$ we define the* variation *of $\mu$ on $B$ by*

$$|\mu|(B) = \sup\Big\{\sum_{i \in \mathbf{N}} |\mu(B_i)| : \ B = \bigcup_i B_i\Big\}.$$

*The set function $|\mu|$ is a positive measure on $\Omega$.*

**Definition 1.6** *The* support *of $\mu \in \mathcal{M}(\Omega; \mathbf{R}^N)$ is defined as*

$$\operatorname{spt} \mu = \Big\{x \in \Omega : \ |\mu|(B_\rho(x)) > 0 \ \textit{ for all } B_\rho(x) \subset \Omega\Big\}.$$

**Theorem 1.7** *Every measure $\mu \in \mathcal{M}_+(\Omega)$ is* regular; *i.e.,*

$$\mu(B) = \inf\Big\{\mu(A) : B \subset A, \ A \ \textit{open}\Big\}, \tag{1.1}$$

$$\mu(B) = \sup\Big\{\mu(C) : C \subset B, \ C \ \textit{closed}\Big\} \tag{1.2}$$

*for all $B \in \mathcal{B}(\Omega)$.*

**Remark 1.8** By approximating closed sets with compact sets we also have

$$\mu(B) = \sup\Big\{\mu(K) : K \subset B, \ K \text{ compact}\Big\}.$$

**Definition 1.9** *Let $\mu \in \mathcal{M}_+(\Omega)$ and $\lambda \in \mathcal{M}(\Omega; \mathbf{R}^N)$. We say that $\lambda$ is* absolutely continuous *with respect to $\mu$ (and we write $\lambda << \mu$) if $\lambda(B) = 0$ for every $B \in \mathcal{B}(\Omega)$ with $\mu(B) = 0$.*

*We say that $\lambda$ is* singular *with respect to $\mu$ if there exists a set $E \in \mathcal{B}(\Omega)$ such that $\mu(E) = 0$ and $\lambda(B) = 0$ for all $B \in \mathcal{B}(\Omega)$ with $B \cap E = \emptyset$ (in this case we say that $\lambda$ is* concentrated *on $E$).*

**Remark 1.10** If $f \in L^1(\Omega, \mu; \mathbf{R}^N)$ and $\mu \in \mathcal{M}(\Omega)$ then we define the measure $f\mu \in \mathcal{M}(\Omega; \mathbf{R}^N)$ by

$$f\mu(B) = \int_B f \, d\mu.$$

We have that $f\mu << |\mu|$. Moreover, $|f\mu| = |f||\mu|$.

**Theorem 1.11. (Radon-Nikodym)** *If $\lambda \in \mathcal{M}(\Omega;\mathbf{R}^N)$, and $\mu \in \mathcal{M}_+(\Omega)$, then there exists a function $f \in L^1(\Omega,\mu;\mathbf{R}^N)$ and a measure $\lambda^s$, singular with respect to $\mu$, such that*

$$\lambda = f\mu + \lambda^s.$$

*This will be called the* Radon-Nikodym decomposition *of $\lambda$ with respect to $\mu$.*

**Remark 1.12** From the theorem above we get:

(a) If $\lambda << \mu$ then $\lambda = f\mu$ for some $f \in L^1(\Omega,\mu;\mathbf{R}^N)$;

(b) Since $\lambda << |\lambda|$ there exists $\nu \in L^1(\Omega,|\lambda|;\mathbf{R}^N)$ such that $\lambda = \nu|\lambda|$. As $|\lambda| = |\nu|\lambda|| = |\nu||\lambda|$ we get that $|\nu| = 1$ $\mu$-a.e. on $\Omega$;

(c) If $\mu \in \mathcal{M}(\Omega)$, we can write $\mu = \mu^+ - \mu^-$, where $\mu^\pm = \nu^\pm|\mu| \in \mathcal{M}_+(\Omega)$ ($\nu^\pm$ denotes the positive/negative part of $\nu$).

The density $f$ in Theorem 1.11 is characterized by the following theorem.

**Theorem 1.13. (Besicovitch Derivation Theorem)** *Let $\mu$, $\lambda$ and $f$ be as in Theorem 1.11. Then for $\mu$-almost all $x \in \operatorname{spt}\mu$ there exists the limit*

$$\frac{d\lambda}{d\mu}(x) = \lim_{\rho\to 0+} \frac{\lambda(B_\rho(x))}{\mu(B_\rho(x))},$$

*and $f(x) = \frac{d\lambda}{d\mu}(x)$ for $\mu$-almost all $x \in \operatorname{spt}\mu$.*

As a corollary we obtain the following useful proposition.

**Corollary 1.14** *Let $\mu \in \mathcal{M}_+(\Omega)$ and $g \in L^1(\Omega,\mu)$. Then*

$$\lim_{\rho\to 0+} \frac{1}{\mu(B_\rho(x))} \int_{B_\rho(x)} |g(x) - g(y)|\,d\mu(y) = 0 \tag{1.3}$$

*$\mu$-a.e. on $\operatorname{spt}\mu$.*

**Definition 1.15** *Let $\mu = \mathcal{L}_n$ and let $g \in L^1(\Omega)$. Then each point $x$ such that (1.3) holds is called a* Lebesgue point *for $g$. Note that the set of the Lebesgue points of $g$ depends on the particular choice of the representative in the equivalence class of $L^1(\Omega)$; hence, we will always think a particular choice of the representative of $g$ as fixed whenever we consider Lebesgue points.*

### 1.1.1 *Supremum of a family of measures*

From the regularity properties of positive measures we easily obtain the following proposition.

**Proposition 1.16** *Let $\mu : \mathcal{A}(\Omega) \to [0,+\infty)$ be an open-set function superadditive on open sets with disjoint compact closures (i.e., $\mu(A\cup B) \geq \mu(A)+\mu(B)$ for all $A, B \in \mathcal{A}(\Omega)$ with $\overline{A}\cap\overline{B} = \emptyset$, $\overline{A}\cup\overline{B} \subset\subset \Omega$), let $\lambda \in \mathcal{M}_+(\Omega)$, let $\psi_i$ be positive Borel functions such that $\mu(A) \geq \int_A \psi_i\,d\lambda$ for all $A \in \mathcal{A}(\Omega)$ and let $\psi(x) = \sup_i \psi_i(x)$. Then $\mu(A) \geq \int_A \psi\,d\lambda$ for all $A \in \mathcal{A}(\Omega)$.*

**Proof** We have, by the regularity of the measures $\psi_i\lambda$,

$$\int_A \psi\, d\lambda = \sup\Big\{\sum_{i=1}^k \int_{B_i} \psi_i\, d\lambda : (B_i) \text{ Borel partition of } A,\ k \in \mathbf{N}\Big\}$$

$$= \sup\Big\{\sum_{i=1}^k \int_{K_i} \psi_i\, d\lambda : (K_i) \text{ disjoint compact subsets of } A,\ k \in \mathbf{N}\Big\}$$

$$= \sup\Big\{\sum_{i=1}^k \int_{A_i} \psi_i\, d\lambda : (A_i) \text{ disjoint open subsets of } A,\ k \in \mathbf{N}\Big\} \leq \mu(A),$$

that is, the thesis. □

## 1.2 Construction of measures. Hausdorff measures

### 1.2.1 *Carathéodory's construction*

We can construct positive measures from set functions, following a procedure due to Carathéodory, which we briefly recall.

**Definition 1.17** *A set function* $\lambda : \mathcal{P}(\Omega) \to [0, +\infty]$ *is an* outer measure *if* $\lambda(\emptyset) = 0$ *and is* countably subadditive; *i.e.,*

$$\lambda(A) \leq \sum_{i\in\mathbf{N}} \lambda(A_i) \qquad \textit{if} \qquad A \subseteq \bigcup_{i\in\mathbf{N}} A_i\,.$$

*We say that a set* $M$ *is* measurable *for* $\lambda$ *if*

$$\lambda(A) = \lambda(A \cup M) + \lambda(A \setminus M)$$

*for all* $A \subseteq \Omega$.

**Theorem 1.18. (Carathéodory's Construction)** *If* $\lambda$ *is an outer measure then the family* $\mathcal{M}_\lambda$ *of all measurable sets for* $\lambda$ *is a* $\sigma$*-algebra and* $\lambda_{|\mathcal{M}_\lambda}$ *is countably additive.*

**Remark 1.19** By the previous theorem, if $\mathcal{B}(\Omega) \subseteq \mathcal{M}_\lambda$ and $\lambda(\Omega) < +\infty$ then $\mu = \lambda_{|\mathcal{B}(\Omega)} \in \mathcal{M}_+(\Omega)$.

To check that an outer measure generates a measure by the construction above, we have to prove that Borel sets are measurable. The following proposition provides an easy criterion for measurability.

**Proposition 1.20. (Carathéodory's Criterion)** *Let* $\lambda$ *be an outer measure. Then Borel sets are measurable for* $\lambda$ *if and only if*

$$\operatorname{dist}(A, B) > 0 \qquad \Longrightarrow \qquad \lambda(A) + \lambda(B) = \lambda(A \cup B) \tag{1.4}$$

*for all* $A, B \subseteq \Omega$, *where* $\operatorname{dist}(A, B) = \inf\{|x - y| : x \in A,\ y \in B\}$.

### 1.2.2 *Hausdorff measures*

We apply Carathéodory's construction to define the measures which will be most of use in the sequel

**Definition 1.21** *Let $\alpha \geq 0$ and $\delta > 0$. For all $E \subset \mathbf{R}^n$ we define the* pre-Hausdorff measure $\mathcal{H}^\alpha_\delta$ *of $E$ as*

$$\mathcal{H}^\alpha_\delta(E) = \frac{\omega_\alpha}{2^\alpha} \inf\Big\{\sum_{i\in\mathbf{N}} (\operatorname{diam} E_i)^\alpha : \ \operatorname{diam} E_i \leq \delta,\ E \subseteq \bigcup_{i\in\mathbf{N}} E_i\Big\},$$

*where $\omega_\alpha = \pi^{\alpha/2}/\Gamma(\alpha/2+1)$, and $\Gamma(\alpha) = \int_0^{+\infty} s^{\alpha-1}e^{-s}\,ds$ is the Euler function, which coincides with the Lebesgue measure of the unit ball in $\mathbf{R}^\alpha$ if $\alpha$ is integer. Note that $\mathcal{H}^\alpha_\delta(E)$ is decreasing in $\delta$.*

*The $\alpha$-dimensional* Hausdorff measure *of $E$ is defined by*

$$\mathcal{H}^\alpha(E) = \sup_{\delta>0} \mathcal{H}^\alpha_\delta(E) = \lim_{\delta\to 0} \mathcal{H}^\alpha_\delta(E).$$

**Proposition 1.22** *$\mathcal{H}^\alpha$ is countably additive on $\mathcal{B}(\Omega)$.*

**Proof** Clearly, $\mathcal{H}^\alpha_\delta$ is countably subadditive on $\mathcal{P}(\mathbf{R}^n)$, hence such is also $\mathcal{H}^\alpha$. It now suffices to remark that $\mathcal{H}^\alpha_\delta(A \cup B) = \mathcal{H}^\alpha_\delta(A) + \mathcal{H}^\alpha_\delta(B)$ whenever $\operatorname{dist}(A, B) > 2\delta$. Letting $\delta \to 0$ we see that we can apply Theorem 1.18. □

**Remark 1.23** (a) $\mathcal{H}^\alpha$ is the null measure if $\alpha > n$;

(b) $\mathcal{H}^0 = \#$ the counting measures, whose value $\#(E)$ is the cardinality of $E$ if $E$ is a finite set, and $+\infty$ otherwise;

(c) If $f : \mathbf{R}^n \to \mathbf{R}^m$ is Lipschitz continuous with constant $\Lambda$ then $\mathcal{H}^\alpha(f(E)) \leq \Lambda^\alpha \mathcal{H}^\alpha(E)$ (note that we use the same notation for the two Hausdorff measures on $\mathbf{R}^n$ and $\mathbf{R}^m$). In particular $\mathcal{H}^\alpha(z + \Lambda E) = \Lambda^\alpha \mathcal{H}^\alpha(E)$;

(d) If $\alpha > \beta \geq 0$ then if $\mathcal{H}^\alpha(E) > 0$ then $\mathcal{H}^\beta(E) = +\infty$.

**Remark 1.24** If $\alpha < n$ then $\mathcal{H}^\alpha$ is not a measure, since $\mathcal{H}^\alpha(\Omega) = +\infty$ for all non-empty open sets $\Omega$. If $\mathcal{H}^\alpha(B) < +\infty$ for some $B \in \mathcal{B}(\Omega)$ then $\mathcal{H}^\alpha \llcorner B \in \mathcal{M}_+(\Omega)$. We will always use measures of this form. It can be proven, moreover, that $\mathcal{H}^n = \mathcal{L}_n$ in $\mathbf{R}^n$.

**Definition 1.25** *We define the* Hausdorff dimension *of a set $E \subseteq \mathbf{R}^n$ as*

$$\dim_{\mathcal{H}}(E) = \inf\{\alpha \geq 0 : \ \mathcal{H}^\alpha(E) = 0\},$$

*which is well-defined by Remark* 1.23(d).

### 1.2.3 *The De Giorgi and Letta measure criterion*

Another way to construct positive measures is by first defining a set function on open sets, as described in the following criterion (for a proof see [DM] or [BDF]).

**Proposition 1.26. (De Giorgi and Letta Criterion)** *Let the open-set function $\lambda : \mathcal{A}(\Omega) \to [0, +\infty)$ satisfy*

(ii) $\lambda(\emptyset) = 0$, $\lambda(A) \le \lambda(B)$ *if* $A \subseteq B$ *($\lambda$ is an* increasing set function*);*
(ii) $\lambda(A \cup B) \le \lambda(A) + \lambda(B)$ *for all* $A, B \in \mathcal{A}(\Omega)$ *($\lambda$ is* subadditive*);*
(iii) $\lambda(A \cup B) \ge \lambda(A) + \lambda(B)$ *for all* $A, B \in \mathcal{A}(\Omega)$ *with* $A \cap B = \emptyset$ *($\lambda$ is* superadditive *on disjoint sets).*
*Then there exists* $\mu \in \mathcal{M}_+(\Omega)$ *such that* $\mu = \lambda$ *on* $\mathcal{A}(\Omega)$.

**Remark 1.27** By regularity, such $\mu$ is given by

$$\mu(B) = \inf\{\lambda(A) : \ A \in \mathcal{A}(\Omega), \ B \subseteq A\}$$

for all $B \in \mathcal{B}(\Omega)$. Note that $\lambda$ satisfies condition (1.4) of Proposition 1.20 on open sets.

## 1.3 Weak convergence of measures

Measures can be identified as elements of the dual of the space of continuous functions vanishing on $\partial\Omega$. Hence, they inherit a notion of weak* convergence which will be useful in the sequel.

**Definition 1.28** *We define the set* $C_0(\Omega; \mathbf{R}^N)$ *as the closure of* $C_c^\infty(\Omega; \mathbf{R}^N)$ *in the uniform topology. It is a separable Banach space if equipped with the* $\|\cdot\|_\infty$ *norm.*

**Remark 1.29** For all $\mu \in \mathcal{M}(\Omega; \mathbf{R}^N)$ we define

$$L_\mu(\phi) = \int_\Omega \phi \, d\mu =: \sum_{i=1}^N \int_\Omega \phi_i d\mu_i \tag{1.5}$$

if $\phi_i \in L^1(\Omega, \mu_i)$. The functional $L_\mu$ is linear and continuous on $C_0(\Omega; \mathbf{R}^N)$.

**Theorem 1.30. (Riesz's Theorem)** *The map* $\mu \mapsto L_\mu$ *defined in* (1.5) *is a bijection between* $\mathcal{M}(\Omega; \mathbf{R}^N)$ *and* $(C_0(\Omega; \mathbf{R}^N))'$.

**Remark 1.31** We have $\|L_\mu\| = |\mu|(\Omega)$. In fact,

$$\begin{aligned}\|L_\mu\| &= \sup\Big\{\int_\Omega \phi d\mu : \ \phi \in C_0(\Omega; \mathbf{R}^N), \ |\phi| \le 1\Big\} \\ &= \sup\Big\{\int_\Omega \langle \phi, \nu\rangle d|\mu| : \ \phi \in C_0(\Omega; \mathbf{R}^N), \ |\phi| \le 1\Big\} \\ &= \int_\Omega \langle \nu, \nu\rangle d|\mu| = |\mu|(\Omega),\end{aligned}$$

since using Lusin's Theorem we can approximate $\nu$ by functions in $C_0(\Omega; \mathbf{R}^N)$.

**Definition 1.32** *We say that a sequence* $(\mu_j) \subset \mathcal{M}(\Omega; \mathbf{R}^N)$ converges weakly *to* $\mu$ *(and we write* $\mu_j \rightharpoonup \mu$*) if* $L_{\mu_j} \rightharpoonup^* L_\mu$ *in the weak* topology of* $(C_0(\Omega; \mathbf{R}^N))'$*; i.e.,*

$$\lim_{j \to +\infty} \int_\Omega \phi d\mu_j = \int_\Omega \phi d\mu$$

*for all* $\phi \in C_0(\Omega; \mathbf{R}^N)$.

**Remark 1.33** By the Banach-Steinhaus Theorem we have that if $\mu_j \rightharpoonup \mu$ then $\sup_j |\mu_j|(\Omega) < +\infty$. Note, moreover, that by the lower semicontinuity of the dual norm with respect to weak* convergence we have that $\mu \mapsto |\mu|(\Omega)$ is weakly lower semicontinuous; i.e., $|\mu|(\Omega) \leq \liminf_j |\mu_j|(\Omega)$ if $\mu_j \rightharpoonup \mu$.

**Theorem 1.34. (Weak Compactness)** *Let $(\mu_j)$ be a sequence in $\mathcal{M}(\Omega;\mathbf{R}^N)$ with $\sup_j |\mu_j|(\Omega) < +\infty$. Then there exists a subsequence of $(\mu_j)$ weakly converging to some $\mu \in \mathcal{M}(\Omega;\mathbf{R}^N)$.*

### 1.3.1 *Weak convergence of measures as set functions*

**Proposition 1.35** *Let $(\mu_j) \subset \mathcal{M}_+(\Omega)$ and $\mu_j \rightharpoonup \mu$. Then we have*

$$\mu(A) \leq \liminf_{j\to+\infty} \mu_j(A) \tag{1.6}$$

*for all open sets $A \subset \Omega$*

$$\mu(K) \geq \limsup_{j\to+\infty} \mu_j(K) \tag{1.7}$$

*for all compact sets $K \subset \Omega$*

**Proposition 1.36** *If $(\mu_j) \subset \mathcal{M}(\Omega;\mathbf{R}^N)$, $\mu_j \rightharpoonup \mu$ and $|\mu_j| \rightharpoonup \sigma$ then $|\mu| \leq \sigma$. Moreover*

$$\mu_j(B) \to \mu(B) \tag{1.8}$$

*for all $B \in \mathcal{B}(\Omega)$, $B \subset\subset \Omega$ such that $\sigma(\partial B) = 0$*

**Proof** For all $A \in \mathcal{A}(\Omega)$ and $\phi \in C_0(A;\mathbf{R}^N)$, from $|\int_A \phi d\mu_j| \leq \int_A |\phi| d|\mu_j|$ we deduce that $|\int_A \phi d\mu| \leq \int_A |\phi| d\sigma$, and hence $|\mu|(A) \leq \sigma(A)$. By (1.1) we obtain $|\mu| \leq \sigma$ on all $\mathcal{B}(\Omega)$.

If $\phi \in C_c(\Omega)$ with $0 \leq \phi \leq 1$ and $\phi = 1$ on $B$ then set $K = \operatorname{spt}\phi \setminus \operatorname{int} B$. We have

$$\begin{aligned}|\mu(B) - \mu_j(B)| &= \Big|\int_\Omega \phi(d\mu - d\mu_j) + \int_{\operatorname{spt}\phi\setminus B} \phi(d\mu - d\mu_j)\Big| \\ &\leq \Big|\int_\Omega \phi(d\mu - d\mu_j)\Big| + |\mu_j|(K) + \sigma(K).\end{aligned}$$

Hence, recalling (1.7), we get $\limsup_j |\mu(B) - \mu_j(B)| \leq 2\sigma(K)$, and, by the arbitrariness of $\phi$ as above, $\limsup_j |\mu(B) - \mu_j(B)| \leq 2\sigma(\partial B) = 0$. □

**Definition 1.37** *A family $\mathcal{R} \subset \mathcal{A}(\Omega)$ is* rich *if for every pair $A' \subset\subset A \in \mathcal{A}(\Omega)$ there exists a family $(A_t)_{0<t<1} \subset \mathcal{A}(\Omega)$ such that $A' \subset A_t \subset\subset A_s \subset A$ for $0 < t < s < 1$ and $A_t \in \mathcal{R}$ for a.a. $t \in (0,1)$.*

**Remark 1.38** Let $\sigma \in \mathcal{M}_+(\Omega)$; then the family of open sets such that $\sigma(\partial A) = 0$ is rich. In fact, given $A' \subset\subset A$ set

$$A_t = \{x \in A : \operatorname{dist}(x, A') < t \operatorname{dist}(A', \mathbf{R}^n \setminus A)\}.$$

Then $\mu(\partial A_t) = 0$ except for at most a countable set of $t \in (0,1)$ since $\partial A_s \cup \partial A_t = \emptyset$ and $\sigma(\bigcup_{0<t<1} \partial A_t) \leq \sigma(\Omega) < +\infty$.

**Proposition 1.39** *Let $\mu_j, \mu \in \mathcal{M}(\Omega; \mathbf{R}^N)$. If $\sup_j |\mu_j|(\Omega) < +\infty$ and there exists a rich family $\mathcal{R}$ of open sets such that $\mu_j(A) \to \mu(A)$ for $A \in \mathcal{R}$ then $\mu_j \rightharpoonup \mu$.*

**Proof** Up to subsequences $\mu_j \rightharpoonup \lambda$ and $|\mu_j| \rightharpoonup \alpha$. We have to show that $\lambda = \mu$. Using the argument in Remark 1.38 we see that the family $\mathcal{R}' = \{A \in \mathcal{R} : \alpha(\partial A) = 0\}$ is still rich. Moreover, by Proposition 1.36 $\mu_j(A) \to \lambda(A)$, so that $\mu(A) = \lambda(A)$, if $A \in \mathcal{R}'$. Hence, if $K \subset \Omega$ is compact

$$\begin{aligned}\mu(K) &= \inf\{\mu(A) : A \in \mathcal{A}(\Omega), K \subset A\} = \inf\{\mu(A) : A \in \mathcal{R}', K \subset A\} \\ &= \inf\{\lambda(A) : A \in \mathcal{R}', K \subset A\} = \inf\{\lambda(A) : A \in \mathcal{A}(\Omega), K \subset A\} = \lambda(K).\end{aligned}$$

By Remark 1.8 we get $\mu = \lambda$ on the whole $\mathcal{B}(\Omega)$. □

### 1.3.2 *Reshetnyak's Theorem*

We end this section by stating a classical theorem by Reshetnyak, which deals with lower semicontinuity and continuity properties of integral functionals defined on spaces of measures.

If $\varphi : \mathbf{R}^N \to [0, +\infty)$ is a continuous function positively homogeneous of degree one, we use the notation

$$\int_\Omega \varphi(\mu) = \int_\Omega \varphi\Big(\frac{d\mu}{d|\mu|}\Big)\, d|\mu| \tag{1.9}$$

for $\mu \in \mathcal{M}(\Omega; \mathbf{R}^N)$.

**Theorem 1.40** *Let $\varphi : \mathbf{R}^N \to [0, +\infty)$ satisfy*

*(i) $\varphi$ is convex;*

*(ii) $\varphi$ is positively homogeneous of degree 1;*

*i.e., $\varphi(z_1 + z_2) \le \varphi(z_1) + \varphi(z_2)$ and $\varphi(tz) = t\varphi(z)$ if $t \ge 0$. Then, if $(\mu_j)$ is a sequence in $\mathcal{M}(\Omega; \mathbf{R}^N)$ weakly converging to $\mu$, we have*

$$\int_\Omega \varphi(\mu) \le \liminf_j \int_\Omega \varphi(\mu_j). \tag{1.10}$$

*Moreover,*

$$\int_\Omega \varphi(\mu) = \lim_j \int_\Omega \varphi(\mu_j) \tag{1.11}$$

*if in addition $|\mu_j|(\Omega) \to |\mu|(\Omega)$.*

## 1.4 BV functions

**Definition 1.41** *Let $u \in L^1(\Omega)$. We say that $u$ is a* function of bounded variation *on $\Omega$ if its distributional derivative is a measure; i.e., there exists $\mu \in \mathcal{M}(\Omega; \mathbf{R}^n)$ such that*

$$\int_\Omega u D\phi\, dx = -\int_\Omega \phi\, d\mu$$

*for all* $\phi \in C_c^1(\Omega)$. *The measure* $\mu$ *will be denoted by* $Du$, *and its components by* $D_1u, \dots, D_nu$. *The space of all functions of bounded variation on* $\Omega$ *will be denoted by* $BV(\Omega)$.

*We say that a sequence* $(u_j)$ converges weakly *in* $BV(\Omega)$, *and we write* $u_j \rightharpoonup u$ *in* $BV(\Omega)$, *if it converges in* $L^1(\Omega)$ *and* $\sup_j |Du_j|(\Omega) < +\infty$.

**Remark 1.42** (a) If $u \in W^{1,1}(\Omega)$ then $u \in BV(\Omega)$ and $|Du|(\Omega) = \int_\Omega |\nabla u| dx$.

(b) If $u_j \rightharpoonup u$ in $BV(\Omega)$ then $u \in BV(\Omega)$, and $Du_j \rightharpoonup Du$ as measures. In fact, let $\mu_j = Du_j$; from the condition $\sup_j |\mu_j|(\Omega) < +\infty$ we deduce that, up to subsequences, $\mu_j \rightharpoonup \mu$. Now it suffices to remark that $\int_\Omega u_j D\phi\, dx = -\int_\Omega \phi\, d\mu_j$ passes to the limit, so that $\mu = Du$.

(c) If $u_j \to u$ in $L^1(\Omega)$ then $|Du|(\Omega) \le \liminf_j |Du_j|(\Omega)$. It suffices to remark that it is not restrictive to suppose $u_j \rightharpoonup u$ in $BV(\Omega)$, and apply the weak lower semicontinuity of the variation.

An approximation by convolution argument proves the following proposition.

**Proposition 1.43** *If* $u \in BV(\Omega)$ *then there exists a sequence* $(u_j)$ *of* $C^\infty$-*functions such that* $u_j \to u$ *in* $L^1(\Omega)$ *and* $|Du_j|(\Omega) \to |Du|(\Omega)$.

**Theorem 1.44** *The following statements are equivalent:*

(i) $u \in BV(\Omega)$;

(ii) $u \in L^1(\Omega)$ *and the* total variation *of* $u$ *on* $\Omega$

$$\sup\Big\{\int_\Omega u \operatorname{div} g\, dx : \ g \in C_c^1(\Omega; \mathbf{R}^n), \ |g| \le 1\Big\} \tag{1.12}$$

*is finite;*

(iii) *there exists a sequence* $(u_j)$ *of* $C^\infty$ *functions such that* $u_j \to u$ *in* $L^1(\Omega)$ *and* $\limsup_j \int_\Omega |\nabla u_j| dx < +\infty$.

**Proof** Let $M$ be the supremum in (1.12). We first show that (i) is equivalent to (ii). One implication is trivial, since if (i) holds then

$$M \le \sup\Big\{\Big|\int_\Omega g\, d\mu\Big| : \ g \in C_0(\Omega; \mathbf{R}^n), \ |g| \le 1\Big\} = |Du|(\Omega).$$

On the other hand, if $L(g) = \int_\Omega u \operatorname{div} g\, dx$ then we get $|L(g)| \le M\|g\|_\infty$ on $C_c^1(\Omega; \mathbf{R}^n)$. Hence, $L$ can be prolonged to an element of $C_0(\Omega; \mathbf{R}^n)$. By Riesz's Theorem $L(g) = -\int_\Omega g\, d\mu$ for some $\mu \in \mathcal{M}(\Omega; \mathbf{R}^n)$; i.e., $u \in BV(\Omega)$. Note that $|\mu|(\Omega) \le M$.

As for the equivalence between (i) and (iii), note that one implication follows from Proposition 1.43. Conversely, if (iii) holds then (i) follows from Remark 1.42(b). □

**Remark 1.45** From the proof of the previous theorem we get that $|Du|(\Omega)$ coincides with the total variation of $u$ on $\Omega$. From Proposition 1.43 and Remark 1.42(c) we have also a variational characterization:

$$|Du|(\Omega) = \inf\Big\{\liminf_j \int_\Omega |\nabla u_j| dx : \ u_j \to u \text{ in } L^1(\Omega), \ u_j \in C^\infty(\Omega)\Big\}.$$

Formula (1.12) extends the definition of $|Du|(\Omega)$ to all $u \in L^1_{\text{loc}}(\Omega)$. Note that $u \mapsto |Du|(\Omega)$ is lower semicontinuous with respect to the $L^1_{\text{loc}}(\Omega)$ convergence, as it is the supremum of a family of continuous functionals.

**Theorem 1.46** *If* $(u_j) \subset BV(\Omega)$ *and* $\sup_j (\|u_j\|_{L^1(\Omega)} + |Du_j|(\Omega)) < +\infty$ *then there exists a subsequence converging in* $L^1_{\text{loc}}(\Omega)$ *to some* $u \in BV(\Omega)$.

**Remark 1.47** If $\Omega$ has a regular boundary, then it can be proven that $BV(\Omega) \subset L^q(\Omega)$ if $q \leq n/(n-1)$, with compact immersion if $q < n/(n-1)$.

### 1.4.1 *BV functions of one variable*

If $u \in BV(a,b)$ then it can be easily seen that there exits a constant $c$ such that

$$u(x) = c + Du(a,x),$$

by differentiating both sides of this equality in the sense of distributions. We deduce then that we can write $u$ as the difference of two increasing functions:

$$u(x) = c + (Du)^+(a,x) - (Du)^-(a,x),$$

where $(Du)^\pm = \nu^\pm |Du|$. Moreover, the right-hand side and left-hand side limits

$$u(t+) = \lim_{h \to 0+} \frac{1}{h} \int_t^{t+h} u(\tau)\, d\tau,$$

$$u(t-) = \lim_{h \to 0+} \frac{1}{h} \int_{t-h}^{t} u(\tau)\, d\tau$$

exist at all $t \in [a,b)$, and $t \in (a,b]$, respectively. Finally, $u(t+) = u(t-)$ if $|Du|(\{t\}) = 0$.

## 1.5 Sets of finite perimeter

**Definition 1.48** *We say that a set* $E \subset \mathbf{R}^n$ *is a* set of finite perimeter in $\Omega$ *if* $\chi_E \in BV(\Omega)$*; that is, if*

$$\sup\Big\{\int_E \operatorname{div} g\, dx : \ g \in C_0^1(\Omega; \mathbf{R}^n), \ |g| \leq 1\Big\} < +\infty.$$

*The quantity* $|D\chi_E|(\Omega)$ *is called the* perimeter of $E$ in $\Omega$.

**Remark 1.49** As $\mathcal{H}^{n-1} = \mathcal{L}_{n-1}$ on $\mathbf{R}^{n-1}$, it can be easily seen that if $E$ is a polyhedron then $|D\chi_E|(\Omega) = \mathcal{H}^{n-1}(\partial E \cap \Omega)$. This equality easily extends by approximation if $\partial E$ is piecewise $C^1$.

**Remark 1.50** If $E$ is of finite perimeter in $\Omega$ and $C \subset \Omega$ satisfies $\mathcal{H}^{n-1}(C) = 0$ then $|D\chi_E|(C) = 0$.

**Theorem 1.51. (Fleming-Rishel coarea formula)** *Let $u \in L^1(\Omega)$; then*

$$|Du|(\Omega) = \int_{-\infty}^{+\infty} |D\chi_{\{u>t\}}|(\Omega)\,dt. \tag{1.13}$$

**Proof** Note that we can write $u(x) = \int_0^{+\infty} \chi_{\{u>t\}}(x)\,dt - \int_{-\infty}^0 (1-\chi_{\{u>t\}}(x))\,dt$. Using this identity, as $\int_\Omega \operatorname{div} g\,dx = 0$, we have

$$\int_\Omega u \operatorname{div} g\,dx = \int_0^{+\infty} \Big( \int_\Omega \chi_{\{u>t\}} \operatorname{div} g\,dx \Big)\,dt \le \int_0^{+\infty} |D\chi_{\{u>t\}}|(\Omega)\,dt\,.$$

The converse inequality will be obtained by a double approximation procedure and using the lower semicontinuity of the total variation. First, let $u$ be a continuous piecewise affine function such that we can write $\Omega = \bigcup_{i=1}^N \Omega_i \cup N$, with $\Omega_i$ disjoint open sets and $\mathcal{H}^{n-1}(N) = 0$, and $u(x) = \langle \xi_i, x\rangle + c_i$ on $\Omega_i$. In this case it can be easily seen, using Remark 1.49 that

$$\int_{-\infty}^{+\infty} |D\chi_{\{u>t\}}|(\Omega_i)\,dt = \int_{\Omega_i} |\nabla u|\,dx = |\xi_i||\Omega_i|\,.$$

Since $\partial\{u > t\}$ is piecewise $C^1$ for all $t$, and $\mathcal{H}^{n-1}(N \cap \{u = t\}) = 0$ for almost all $t$, we have

$$\int_{-\infty}^{+\infty} |D\chi_{\{u>t\}}|(\Omega)\,dt = \int_{-\infty}^{+\infty} \mathcal{H}^{n-1}(\Omega \cap \{u > t\})\,dt$$

$$= \int_{-\infty}^{+\infty} \sum_{i=1}^N \mathcal{H}^{n-1}(\Omega_i \cap \{u > t\})\,dt = \sum_{i=1}^N |\xi_i||\Omega_i| = |Du|(\Omega)\,.$$

If $u \in C^\infty(\Omega)$ we approximate it in the $W^{1,1}_{\mathrm{loc}}(\Omega)$-convergence by continuous piecewise affine functions $(u_j)$, and remark that $\chi_{\{u_j>t\}} \to \chi_{\{u>t\}}$ if $|\{u = t\}| = 0$. By the lower semicontinuity of the total variation and Fatou's Lemma, we get for all $A \subset\subset \Omega$

$$\int_{-\infty}^{+\infty} |D\chi_{\{u>t\}}|(A)\,dt \le \int_{-\infty}^{+\infty} \liminf_{j\to+\infty} |D\chi_{\{u_j>t\}}|(A)\,dt$$

$$\le \liminf_{j\to+\infty} \int_{-\infty}^{+\infty} |D\chi_{\{u_j>t\}}|(A)\,dt = \liminf_{j\to+\infty} |Du_j|(A) = |Du|(A) \le |Du|(\Omega)\,.$$

By the arbitrariness of $A$ we have the required inequality. Eventually, we can repeat the argument with $u \in BV(\Omega)$, and a sequence $(u_j) \subset C^\infty(\Omega)$ approximating $u$ in $L^1(\Omega)$, such that $|Du_j|(\Omega) \to |Du|(\Omega)$. □

**Remark 1.52** Theorem 1.51 can be stated with each partial derivative $D_i$ in place of $D$, and with the measure derivative in place of the total variation; i.e.,

$$Du(\Omega) = \int_{-\infty}^{+\infty} D\chi_{\{u>t\}}(\Omega)\, dt$$

for $u \in BV(\Omega)$.

**Remark 1.53** From Remark 1.50 and the Fleming-Rishel formula we get that if $u \in BV(\Omega)$ and $\mathcal{H}^{n-1}(C) = 0$, then $|Du|(C) = 0$.

## 1.6 Structure of the sets of finite perimeter

Before stating De Giorgi's theorem which describes the geometrical properties of the sets of finite perimeter (for a proof we refer e.g. to Ambrosio [A5]), we introduce the necessary measure-theoretical objects.

**Definition 1.54** *Let $E$ be any Borel subset of $\mathbf{R}^n$. We say that $x$ is a* point of density $t \in [0,1]$ *if there exists the limit*

$$\lim_{\rho \to 0+} \frac{|E \cap B_\rho(x)|}{\omega_n \rho^n} = t\,.$$

*The set of all points of density $t$ will be denoted by $E_t$.*

*Let $E$ be of finite perimeter in $\Omega$. The De Giorgi's* reduced boundary *of $E$, denoted by $\partial^* E$, is defined by*

$$\partial^* E = \left\{ x \in \operatorname{spt} |D\chi_E| : \ \exists \lim_{\rho \to 0+} \frac{D\chi_E(B_\rho(x))}{|D\chi_E(B_\rho(x))|} =: \nu(x) \in S^{n-1} \right\}.$$

*The function $\nu : \partial^* E \to S^{n-1}$ is called the* interior normal *to $E$.*

*A set $S \subset \mathbf{R}^n$ is* rectifiable *if there exists a countable family $(\Gamma_i)$ of graphs of Lipschitz functions of $(n-1)$ variables such that $\mathcal{H}^{n-1}(S \setminus \bigcup_{i=1}^\infty \Gamma_i) = 0$.*

**Theorem 1.55. (De Giorgi's Rectifiability Theorem)** *Let $E \subset \mathbf{R}^n$ be a set of finite perimeter in $\Omega$. Then*

(i) *$\partial^* E$ is rectifiable;*

(ii) *$|D\chi_E|(B) = \mathcal{H}^{n-1}(B \cap \partial^* E)$. In particular $\mathcal{H}^{n-1}(\partial^* E) < +\infty$;*

(iii) *$\mathcal{H}^{n-1}(\Omega \setminus (E_0 \cup \partial^* E \cup E_1)) = \mathcal{H}^{n-1}(\Omega \cap (\partial^* E \Delta E_{1/2})) = 0$;*

(iv) *the* generalized Gauss-Green formula

$$\int_E \operatorname{div} g\, dx = - \int_{\partial^* E} \langle \nu, g \rangle \, d\mathcal{H}^{n-1} \tag{1.14}$$

*holds for all $g \in C_c^1(\Omega, \mathbf{R}^n)$; i.e., $D\chi_E = \nu \mathcal{H}^{n-1} \llcorner \partial^* E$.*

**Remark 1.56** Using Theorem 1.55 we can rewrite the Fleming-Rishel formula as

$$|Du|(\Omega) = \int_{-\infty}^{+\infty} \mathcal{H}^{n-1}(\Omega \cap \partial^*\{u > t\})\, dt.$$

Note that this formula holds for all open set $A \subset \Omega$ in place of $\Omega$, and, since both sides define finite measures, we have

$$|Du|(B) = \int_{-\infty}^{+\infty} \mathcal{H}^{n-1}(B \cap \partial^*\{u > t\})\, dt$$

for all Borel sets $B \subset \Omega$, by regularity.

## 1.7 Approximate continuity

**Definition 1.57** *Let $u : \Omega \to \mathbf{R}$ be a Borel function. Let $x \in \Omega$. We define the* approximate upper and lower limits *of $u$ as $y \to x$,*

$$\text{ap-}\limsup_{y \to x} u(y) = \inf\{t : \{u > t\} \text{ has density } 0 \text{ in } x\},$$

$$\text{ap-}\liminf_{y \to x} u(y) = \sup\{t : \{u > t\} \text{ has density } 1 \text{ in } x\},$$

*We define $u^+(x) = \text{ap-}\limsup_{y \to x} u(y)$ and $u^- = \text{ap-}\liminf_{y \to x} u(y)$. We say that $u$ is* approximately continuous *at $x$ if $u^+(x) = u^-(x)$. In this case, we denote the common value by $\tilde{u}(x)$ or $\text{ap-}\lim_{y \to x} u(y)$. Finally, we define the* jump set *of $u$ by*

$$S(u) = \{x \in \Omega : \nexists\ \text{ap-}\lim_{y \to x} u(y)\}, \tag{1.15}$$

*so that $\tilde{u}$ is defined on $\Omega \setminus S(u)$.*

**Proposition 1.58** *Let $u : \Omega \to \mathbf{R}$ be a Borel function. We then have:*

(a) *$x \in \Omega \setminus S(u)$ and $z = \tilde{u}(x)$ if and only if for all $\varepsilon > 0$ $x$ is a point of density* 0 *for $\{|u - z| > \varepsilon\}$;*

(b) *if $u \in L^1(\Omega)$, and $\lim_{\rho \to 0+} \rho^{-n} \int_{B_\rho(x)} |u(y) - z|\, dy = 0$ then $u$ is approximately continuous at $x$ and $\tilde{u}(x) = z$;*

(c) *if $u \in L^\infty(\Omega)$ and $x \notin S(u)$ then $\lim_{\rho \to 0+} \rho^{-n} \int_{B_\rho(x)} |u(y) - \tilde{u}(x)|\, dy = 0$;*

(d) $|S(u)| = 0$.

**Definition 1.59** *Let $u : \Omega \to \mathbf{R}$ be a Borel function, and let $x \in \Omega \setminus S(u)$. We say that $u$ is* approximately differentiable *at $x$ if $\xi \in \mathbf{R}^n$ exists such that*

$$\text{ap-}\lim_{y \to x} \frac{|u(y) - \tilde{u}(x) - \langle \xi, y - x\rangle|}{|y - x|} = 0.$$

*The vector $\xi$ is called the* approximate gradient *of $u$ at $x$, and it is denoted by $\nabla u(x)$. If $n = 1$ we also write $u'$ in place of $\nabla u$.*

## 1.8 Structure of BV functions

**Definition 1.60** *Let $u \in BV(\Omega)$. We define the three measures $D^a u$, $D^j u$ and $D^c u$ as follows. By the Radon-Nikodym Theorem we set $Du = D^a u + D^s u$ where $D^a u << \mathcal{L}_n$ and $D^s u$ is the* singular part *of $Du$ with respect to $\mathcal{L}_n$. $D^a u$ is the* absolutely continuous *part of $Du$ with respect to the Lebesgue measure, $D^j u = Du \llcorner S(u)$ is the* jump part *of $Du$, and $D^c u = D^s u \llcorner (\Omega \setminus S(u))$ is the* Cantor part *of $Du$. We can then write*

$$Du = D^a u + D^j u + D^c u .$$

*Note, moreover, that by Remark* 1.53 *we have $|D^c u|(B) = 0$ for all Borel sets $B$ such that $\mathcal{H}^{n-1}(B) = 0$.*

From the following theorem we obtain a characterization of $D^a u$.

**Theorem 1.61** *Let $u \in BV(\Omega)$, and let $\varphi = d D^a u / d\mathcal{L}_n$ as defined by the Besicovitch Derivation Theorem. Then we have*

$$\lim_{\rho \to 0+} \frac{1}{\rho^n} \int_{B_\rho(x)} \frac{|u(y) - u(x) - \langle \varphi(x), y - x \rangle|}{|x - y|} \, dy = 0$$

*for almost all $x \in \Omega$.*

**Theorem 1.62** *If $u \in BV(\Omega)$ then for almost all $x \in \Omega$ there exists the approximate gradient of $u$, and $\nabla u = d D^a u / d\mathcal{L}_n$.*

**Proof** It is an immediate consequence of the previous theorem, once we recall Proposition 1.58(2), and that $u = \tilde{u}$ a.e. □

We now characterize the jump part of $Du$.

**Theorem 1.63** *If $u \in BV(\Omega)$ then $S(u)$ is rectifiable and we have*

$$D^j u = (u^+ - u^-) \nu_u \mathcal{H}^{n-1} \llcorner S(u), \tag{1.16}$$

*where $\nu_u$ is defined by $Du = \nu_u |Du|$; i.e., $D^j u(B) = \int_{B \cap S(u)} (u^+ - u^-) \nu_u \, d\mathcal{H}^{n-1}$.*

**Proof** Set $E(t) = \{u > t\}$, and define $S_t = \Omega \setminus \big(E(t)_0 \cup E(t)_1\big)$, where $E_s$ denotes the set of points of density $s$ for $E$. By Theorem 1.55 each $S_t$ is a rectifiable set whenever $E(t)$ is of finite perimeter in $\Omega$, which we can suppose holds for $t$ in a countable dense set $D \subset \mathbf{R}$ by Fleming-Rishel formula. Let $x \in \Omega \setminus S_t$ for all $t \in D$, and set $z = \sup\{t \in D : x \in E(t)_1\}$. By the definition of approximate limits it can be immediately checked that $u^-(x) \geq z$, and $u^+(x) \leq z$. Hence $\tilde{u}(x) = z$ and $x \notin S(u)$, so that $S(u) \subset \bigcup_{t \in D} S_t$; in particular $S(u)$ is rectifiable.

Now, from the definition of $u^\pm$ we get that

$$\chi_{(u^-(x), u^+(x))}(t) \leq \chi_{S_t}(x) \leq \chi_{[u^-(x), u^+(x)]}(t).$$

Let $B \subset S(u)$ with $\mathcal{H}^{n-1}(B) < +\infty$. We then have, by the Fleming-Rishel formula and Theorem 1.55,

$$\begin{aligned}|Du|(B) &= \int_{-\infty}^{+\infty} \mathcal{H}^{n-1}(B \cap \partial^* E(t))\,dt = \int_{-\infty}^{+\infty} \mathcal{H}^{n-1}(B \cap S_t)\,dt \\ &= \int_{-\infty}^{+\infty} \int_B \chi_{S_t}(x)\,d\mathcal{H}^{n-1}(x)\,dt = \int_B \int_{-\infty}^{+\infty} \chi_{S_t}(x)\,dt\,d\mathcal{H}^{n-1}(x) \\ &= \int_B (u^+ - u^-)\,d\mathcal{H}^{n-1}(x).\end{aligned}$$

By the arbitrariness of $B$ we have $|Du| \llcorner S(u) = (u^+ - u^-)\mathcal{H}^{n-1} \llcorner S(u)$, so that (1.16) follows by the definition of $\nu_u$. □

**Remark 1.64** Note that from the proof of Theorem 1.63 we see that for $\mathcal{H}^{n-1}$-a.e. $x \in S(u)$ $\nu_u(x)$ is the normal to $\partial^*\{u > t\}$ for $u^-(x) < t < u^+(x)$. From the Implicit Function Theorem it can be immediately checked that if $u, v \in BV(\Omega)$, then $\nu_u = \pm\nu_v$ $\mathcal{H}^{n-1}$-a.e. on $S(u) \cap S(v)$.

**Definition 1.65** *With a slight abuse of notation, given a function $u \in BV(\Omega)$, if no confusion may arise, we denote by $(u^+(x), u^-(x), \nu_u(x))$ ($x \in S(u)$) either the triplet defined above or $(u^-(x), u^+(x), -\nu_u(x))$. With this notation clearly Theorem* 1.63 *still holds. Note that given $u, v \in BV(\Omega)$ we can choose $\nu_u$ and $\nu_v$ such that $\nu_u = \nu_v$ $\mathcal{H}^{n-1}$-a.e. on $S(u) \cap S(v)$ from Remark* 1.64.

The Cantor part of $Du$ satisfies the following property.

**Theorem 1.66** *Let $u \in BV(\Omega)$, and let $B$ be a Borel set with $\mathcal{H}^{n-1}(B) < +\infty$. Then $|D^c u|(B) = 0$.*

**Proof** We have to show that $|Du|(B) = 0$ if $B \subset \Omega \setminus S(u)$ with $\mathcal{H}^{n-1}(B) < +\infty$. Let $\{u > t\}$ have finite perimeter in $\Omega$ and $|D\chi_{\{u>t\}}|(B) > 0$. Since $\{u > t\}$ has density $\frac{1}{2}$ $\mathcal{H}^{n-1}$-a.e. on $\partial^*\{u > t\}$ we have $\tilde{u} = t$ $\mathcal{H}^{n-1}$-a.e. on $B \cap \partial^*\{u > t\}$. Hence $\mathcal{H}^{n-1}(B \cap \{\tilde{u} = t\}) > 0$, so that the set $\{t \in \mathbf{R} : |D\chi_{\{u>t\}}|(B) > 0, |D\chi_{\{u>t\}}|(\Omega) < +\infty\}$ is at most countable. By the Fleming-Rishel formula we then get $|Du|(B) = 0$. □

**Remark 1.67** By using Reshetnyak's Theorem it is possible to prove the weak lower semicontinuity of convex functionals in spaces of measures, and then in spaces of functions of bounded variation. In particular, if $\varphi : [0, +\infty) \to \mathbf{R}$ is an increasing convex function satisfying $\lim_{t\to+\infty} \varphi(t)/t = M$, then the functional defined on $BV(\Omega)$ by

$$F(u) = \int_\Omega \varphi(|\nabla u|)\,dx + M|D^s u|(\Omega)$$

is lower semicontinuous with respect to the $L^1(\Omega)$ convergence.

### 1.8.1 *1-dimensional sections of BV functions*

We show how to recover directional derivatives of $BV$-functions by considering one-dimensional sections. For the proofs we refer to [AFP1].

We first introduce some notation. Let $\xi \in S^{n-1}$, and let $\Pi_\xi = \{y \in \mathbf{R}^n : \langle y, \xi\rangle = 0\}$ be the linear hyperplane orthogonal to $\xi$. If $y \in \Pi_\xi$ and $E \subset \mathbf{R}^n$ we define $E_{\xi,y} = \{t \in \mathbf{R} : \ y + t\xi \in E\}$. Moreover, if $u : \Omega \to \mathbf{R}$ we set $u_{\xi,y} : \Omega_{\xi,y} \to \mathbf{R}$ by $u_{\xi,y}(t) = u(y + t\xi)$.

If $u \in BV(\Omega)$ we see that for fixed $\xi$ the function $u_{\xi,y}$ belongs to $BV(\Omega_{\xi,y})$ for $\mathcal{H}^{n-1}$-a.a. $y \in \Pi_\xi$. Moreover, from the equality

$$\langle Du(B), \xi\rangle = \int_{\Pi_\xi} Du_{\xi,y}(B_{\xi,y})\, d\mathcal{H}^{n-1}(y),$$

valid for all $B$, we deduce the following theorem.

**Theorem 1.68** *Let $u \in BV(\Omega)$, and let $\xi \in S^{n-1}$. Then we have*

$$\langle D^k u(B), \xi\rangle = \int_{\Pi_\xi} D^k u_{\xi,y}(B_{\xi,y})\, d\mathcal{H}^{n-1}(y),$$

*where $k = a, j$ or $c$.*

**Remark 1.69** From the theorem above we deduce that $\langle \nabla u(y+tx), \xi\rangle = u'_{\xi,y}(t)$ for $\mathcal{H}^{n-1}$-a.a. $y \in \Pi_\xi$ and a.a. $t \in \Omega_{\xi,y}$. Moreover, note that $u_{\xi,y}{}^{\pm}(t) = u^{\pm}(y+t\xi)$.

If $g \in L^1(S(u), \mathcal{H}^{n-1})$ then we get

$$\int_{\Pi_\xi} \int_{B_{\xi,y}\cap S(u_{\xi,y})} g_{\xi,y}(t)\, d\mathcal{H}^0(t)\, d\mathcal{H}^{n-1}(y) = \int_{S(u)\cap B} g(x)|\langle \nu_u, \xi\rangle|\, d\mathcal{H}^{n-1}(x),$$

taking into account that $\nu_{u_{\xi,y}}(t) = +1$ or $-1$ according to the cases $u_{\xi,y}(t+) = u_{\xi,y}{}^{+}(t)$ or $u_{\xi,y}(t+) = u_{\xi,y}{}^{-}(t)$.

### 1.8.2 *The chain rule formula*

**Remark 1.70** Let $\phi : \mathbf{R} \to \mathbf{R}$ be a Lipschitz function, and let $u \in BV(\Omega)$. Then the composition $\phi(u)$ still belongs to $BV(\Omega)$, and $|D(\phi(u))| \leq Lip(\phi)|Du|$. In fact, this inequality holds for $C^\infty(\Omega)$ functions, and passes to the limit under the approximation in Proposition 1.43.

**Proposition 1.71** *Let $\phi : \mathbf{R} \to \mathbf{R}$ be a $C^1$ Lipschitz function, and let $u \in BV(\Omega)$. Then*

$$D(\phi(u)) = \phi'(\tilde{u})\,(D^a u + D^c u) + (\phi(u^+) - \phi(u^-))\,\nu_u\, \mathcal{H}^{n-1} \llcorner S(u)\ . \tag{1.17}$$

**Proof** We can write $\phi(t) = (\phi(t) + (Lip(\phi) + 1)t) - (Lip(\phi) + 1)t$; hence, it suffices to prove (1.17) when in addition $\phi' \geq 1$. Let $v = \phi(u)$; by Remark 1.70 $v \in BV(\Omega)$. Moreover, since $\{v > t\} = \{u > \phi^{-1}(t)\}$, we get $S(v) = S(u)$, $v^{\pm} = \phi(u^{\pm})$ and $\nu_v = \nu_u$; i.e., $D^j v = (\phi(u^+) - \phi(u^-))\nu_u \mathcal{H}^{n-1} \llcorner S(u)$.

Let $B \subset \Omega \setminus S(u)$. Then

$$Dv(B) = \int_{-\infty}^{+\infty} D\chi_{\{v>\tau\}}(B)\, d\tau = \int_{-\infty}^{+\infty} D\chi_{\{u>\phi^{-1}(\tau)\}}(B)\, d\tau$$

$$= \int_{-\infty}^{+\infty} D\chi_{\{u>t\}}(B)\,\phi'(t)\,dt = \int_{-\infty}^{+\infty}\int_B \phi'(t)\,dD\chi_{\{u>t\}}\,dt\,.$$

Note that if $x \in \partial^*\{u > t\} \setminus S(u)$ then $\tilde{u}(x) = t$; hence,

$$\begin{aligned} Dv(B) &= \int_{-\infty}^{+\infty}\int_B \phi'(\tilde{u})\,dD\chi_{\{u>t\}}\,dt \\ &= \int_B \phi'(\tilde{u})\int_{-\infty}^{+\infty} dD\chi_{\{u>t\}}\,dt = \int_B \phi'(\tilde{u})\,dDu\,, \end{aligned}$$

which gives $D^a v + D^c v = \phi'(\tilde{u})\,Du \llcorner (\Omega \setminus S(v))$ as required. □

## 1.9 Exercises

*Exercises to Section* 1.1

**Exercise 1.1** Prove that if $\mu \in \mathcal{M}(\Omega; \mathbf{R}^N)$ then $|\mu|(\Omega) < +\infty$.

*Hint*: by contradiction we can suppose that there exists a disjoint sequence $(B_i)$ such that $\mu_1(B_i) > 0$ and $\sum_i \mu_1(B_i) = +\infty$. In this case $\mu(\cup_i B_i)$ is not defined.

**Exercise 1.2** Prove that $|\mu|$ is the least positive measure among all measures $\lambda$ such that $|\mu(B)| \le \lambda(B)$ for all $B \in \mathcal{B}(\Omega)$.

**Exercise 1.3** Prove that $\operatorname{spt}\mu$ is the intersection of all closed subsets $C$ of $\Omega$ such that $\mu(\Omega \setminus C) = 0$.

**Exercise 1.4** Prove Theorem 1.7.

*Hint*: set $\mathcal{G} = \{B \in \mathcal{B}(\Omega) :$ (1.1) holds both for $B$ and $\Omega \setminus B\}$. Check that $\mathcal{G}$ is a $\sigma$-algebra and contains all closed sets.

**Exercise 1.5** Prove that $\lambda << \mu$ if and only if for all $\varepsilon > 0$ there exists $\delta > 0$ such that if $\mu(A) < \delta$ then $|\lambda|(A) < \varepsilon$.

**Exercise 1.6** Prove Corollary 1.14.

*Hint*: consider the countable family of measures $|f(x) - q|\,\mu$, where $q \in \mathbf{Q}$. By Theorem 1.11 there exists a set $N$ with $\mu(N) = 0$ such that

$$\lim_{\rho\to 0+} (\mu(B_\rho(x)))^{-1}\int_{B_\rho(x)} |f(y) - q|\,d\mu(y) = |f(x) - q|$$

for all $q$ if $x \in \operatorname{spt}\mu \setminus N$. Integrating the inequality $|f(x) - f(y)| \le |f(x) - q| + |f(y) - q|$, and taking the infimum among all $q$, deduce the thesis.

*Exercises to Section* 1.2

**Exercise 1.7** Prove Proposition 1.20.

*Hint*: one implication is trivial since open sets separate sets of positive mutual distance. To prove the converse note that it suffices to show that $\lambda(A) \ge \lambda(A \cap$

$C) + \lambda(A \setminus C)$ for $\lambda(A) < +\infty$ and $C$ closed, since $\mathcal{M}_\lambda$ is a $\sigma$-algebra. Note that $\lambda(\{x \in A: 0 < \text{dist}(x, C) < \varepsilon\}) \to 0$ as $\varepsilon \to 0$. Hence

$$\begin{aligned}\lambda(A \cap C) + \lambda(A \setminus C) \le \lambda(A \cap C) &+ \lambda(\{x \in A: \text{dist}(x, C) \ge \varepsilon)\\ &+\lambda(\{x \in A: 0 < \text{dist}(x, C) < \varepsilon\})\\ \le \lambda(A) &+ \lambda(\{x \in A: 0 < \text{dist}(x, C) < \varepsilon\}),\end{aligned}$$

which proves the required inequality as $\varepsilon \to 0$. We have used the fact that $\lambda$ is increasing. with respect to set inclusion.

**Exercise 1.8** Prove statements (a)–(d) in Remark 1.23. They all are easily derived from the definition of Hausdorff pre-measures; in particular note that (d) comes from the inequality $\mathcal{H}^\alpha_\delta(E) \le \delta^{\alpha-\beta}\mathcal{H}^\beta_\delta(E)$.

*Exercises to Section* 1.3

**Exercise 1.9** Prove that $\phi \in C_0(\Omega; \mathbf{R}^N)$ if and only if $\phi \in C(\Omega; \mathbf{R}^N)$ and for all $\varepsilon > 0$ there exists $K \subseteq \Omega$ compact, such that $|\phi| < \varepsilon$ on $\Omega \setminus K$.

**Exercise 1.10** Prove Theorem 1.34.

*Hint*: by a diagonal argument choose $(j_k)$ such that $\int_\Omega \phi d\mu_{j_k}$ converges for all $\phi$ in a countable family $\Phi$ dense in $C_0(\Omega; \mathbf{R}^N)$. Call $L(\phi)$ this limit. The functional $L$ can be extended from $\Phi$ to a linear and continuous functional on $C_0(\Omega; \mathbf{R}^N)$. By Riesz's Theorem $L(\phi) = \int_\Omega \phi d\mu$ for some $\mu$ and for all $\phi \in C_0(\Omega; \mathbf{R}^N)$. Using the density of the family $\Phi$ deduce that $L_h(\phi) \to L(\phi)$ for all $\phi$.

**Exercise 1.11** Prove (1.6) and (1.7).

*Hint*: deduce (1.6) from Remark 1.33. As for (1.7), note that from the equality $\chi_K = \inf\{\phi \in C_0(A): 0 \le \phi \le 1, \ \phi = 1 \text{ on } K\}$ (where $K \subset A$ and $A \subset \Omega$ is open) we have $\limsup_j \mu_j(K) \le \mu(A)$. Then (1.7) follows from the regularity of $\mu$.

**Exercise 1.12** Let $\mu_j = \delta_j$, $\Omega = \mathbf{R}$. Prove that $\mu_j \rightharpoonup 0$.

**Exercise 1.13** Let $\mu_j = \sin(jx)\mathcal{L}_1$, $\Omega = (0, 1)$. Prove that $\mu_j \rightharpoonup 0$

**Exercise 1.14** Find a weakly converging sequence of measures $\mu_j$ such that $|\mu_j|$ does not converge.

**Exercise 1.15** From Proposition 1.35 directly deduce that $\mu_j(B) \to \mu(B)$ for all $B \in \mathcal{B}(\Omega)$ such that $B \subset\subset \Omega$ and $\mu(\partial B) = 0$.

In fact, we get $\mu(\text{int } B) \le \liminf_j \mu_j(\text{int } B) \le \limsup_j \mu_j(\overline{B}) \le \mu(\overline{B}) = \mu(\text{int } B)$.

*Exercises to Section* 1.4

**Exercise 1.16** Show that if $u_j \rightharpoonup u$ in $BV(\Omega)$ then in general we do not have $|Du_j - Du| \to 0$ even if $|Du_j|(\Omega) \to |Du|(\Omega)$.

*Hint*: take $Du \perp Du_j$ for all $j$, and notice that in this case $|Du_j - Du| = |Du_j| + |Du|$.

**Exercise 1.17** Let $u \in BV(\mathbf{R}^n)$ and let $\rho$ be a mollifier. Prove that $|D(u * \rho)|(\mathbf{R}^n) \le |Du|(\mathbf{R}^n)$. Deduce a proof for Proposition 1.43 in the case $\Omega = \mathbf{R}^n$.

**Exercise 1.18** Prove Theorem 1.46

*Hint*: by Proposition 1.43 it is not restrictive to suppose $u_j \in C^\infty(\Omega)$. Let $\rho$ be a mollifier with support equal to $B_1(0)$, and let $A \subset\subset \Omega$. Let $\gamma < \operatorname{dist}(A, \partial\Omega)$ and let $u_j^\gamma = u_j * \rho_\gamma$ be defined on $A$. Note that

$$\int_A |u_j * \rho_\gamma - u_j|dx \le \int_A \int_{\mathbf{R}^n} \int_0^1 |\nabla u_j(x+ty)||y|\rho_\gamma(y)\,dt\,dy\,dx \tag{1.18}$$

$$\le \gamma \|\nabla u_j\|_{L^1(\Omega)}\,.$$

The sequence $(u_j^\gamma)_j$ is equibounded and equi-Lipschitz continuous, so that by Ascoli-Arzelà's Theorem it converges uniformly, up to subsequences. By a diagonal argument we can suppose that $u_j^{1/k} \to w^k$ for all $k \in \mathbf{N}$. By (1.18) we deduce that $w^k$ converges to some $u$ and $u_j \to u$ as well.

*Exercises to Section* 1.5

**Exercise 1.19** If $\Omega$ is a bounded open set and $(E_j)$ is a sequence of sets of finite perimeter in $\Omega$ such that $\sup_j |D\chi_{E_j}|(\Omega) < +\infty$ then there exists a subsequence (not relabeled) and a set of finite perimeter $E$ such that $\chi_{E_j} \to \chi_E$ in $L^1(\Omega)$ and $|D\chi_E|(\Omega) \le \liminf_j |D\chi_{E_j}|(\Omega)$. (Use Theorem 1.46.)

**Exercise 1.20** Prove the existence of sets of finite perimeter in $\Omega$ solutions of the minimum problems

$$\min\Big\{|D\chi_E|(\Omega) :\ |E| = c\Big\}, \qquad \min\Big\{|D\chi_E|(\Omega) + \int_E g\,dx\Big\},$$

where $0 \le c \le |\Omega|$ and $g \in L^1(\Omega)$ (use the previous exercise).

**Exercise 1.21** Prove that if $E$ is a set of finite perimeter in $\Omega$, then there exists a sequence of sets $(E_j)$ with $\Omega \cap \partial E_j$ of class $C^\infty$ such that

$$|E_j \Delta E| \to 0 \qquad \text{and} \qquad |D\chi_{E_j}|(\Omega) \to |D\chi_E|(\Omega)\,.$$

*Hint*: use Proposition 1.43 to approximate $\chi_E$ by a sequence $(u_j)$ of smooth functions. Note that it is not restrictive to suppose that $0 < u_j < 1$ for all $j$. Use the Fleming and Rishel formula and Fatou's Lemma to show that for a set of $t$ of positive measure $\liminf_j |D\chi_{\{u_j>t\}}|(\Omega) \le |D\chi_E|(\Omega)$. Use Sard's Lemma to find such a $t$ with $\partial\{u_j > t\}$ smooth for all $j$. Choose a subsequence such that $\limsup_j |D\chi_{\{u_j>t\}}|(\Omega) \le |D\chi_E|(\Omega)$, and show the converse inequality for the lim inf using the lower semicontinuity of the total variation.

**Exercise 1.22** Prove Remark 1.50.

*Hint*: with fixed $\varepsilon > 0$, let $C \subset \bigcup_{i=1}^N B_{\rho_i}(x_i)$ and $\sum_{i=1}^N \rho_i^{n-1} < \varepsilon$. Choose functions $\varphi_i \in 1 + C_c^\infty(B_{2\rho_i}(x_i))$, extended to 1 outside $B_{2\rho_i}(x_i)$, with $\varphi_i = 0$ on $B_{\rho_i}(x_i)$, $0 \le \varphi_i \le 1$, and $\|D\varphi_i\|_\infty \le 2/\rho_i$. Set $u_\varepsilon = \chi_E\,(\Pi_{i=1}^N \varphi_i)$, so that $u_\varepsilon = 0$ in a neighbourhood of $C$ and $u_\varepsilon = \chi_E$ in $\Omega \setminus \bigcup_{i=1}^N B_{2\rho_i}(x_i)$. Note that

$$|Du_\varepsilon|(\Omega) = |Du_\varepsilon|(\Omega \setminus C) \le |D\chi_E|(\Omega \setminus C) + 2^{n+1}\omega_n \varepsilon.$$

Since $u_\varepsilon \to \chi_E$ in $L^1(\Omega)$, we get

$$|D\chi_E|(\Omega) \le \liminf_{\varepsilon \to 0+} |Du_\varepsilon|(\Omega) \le |D\chi_E|(\Omega \setminus C),$$

so that $|D\chi_E|(C) = 0$.

*Exercises to Section* 1.6

**Exercise 1.23** Show that we may have $|\partial E| > 0$ even if $E$ is a set of finite perimeter. This shows that $\partial^* E$ and $\partial E$ may be substantially different.

*Hint*: choose a dense sequence $(x_i)$ in $\mathbf{R}^2$, and define $E = \bigcup_i B_{\rho_i}(x_i)$ for a suitable choice of $(\rho_i)$.

*Exercises to Section* 1.7

**Exercise 1.24** Prove Proposition 1.58.

*Hint*: (a): if $z = \tilde{u}(x)$ write $\{|u - z| > \varepsilon\}$ as a union $\{u < z - \varepsilon\} \cup \{u > z + \varepsilon\}$. From the definition of $u^\pm(x)$ deduce that $x$ is a point of density 0 for each of the two sets. Conversely, if $x$ is a point of density 0 for both sets and for all $\varepsilon > 0$ deduce that $u_+(x) \le z + \varepsilon$ and $u_-(x) \ge z - \varepsilon$ for all $\varepsilon > 0$, and (1). (b): use the inequality $\varepsilon |\{|u - z| > \varepsilon\} \cap B_\rho(x)| \le \int_{B_\rho(x)} |u(y) - z|\, dy$. (c): use the inequality $\int_{B_\rho(x)} |u(y) - z|\, dy \le \varepsilon\, |B_\rho(x) \setminus \{|u - z| > \varepsilon\}| + 2\|u\|_\infty |B_\rho(x) \cap \{|u - z| > \varepsilon\}|$. (d): if $u \in L^\infty(\Omega)$ use Corollary 1.14 and (3). In the general case write $S(u) = \bigcup_{j \in \mathbf{N}} S(u_j)$ where $u_j(x) = -j \vee (u(x) \wedge j)$.

*Exercises to Section* 1.8

**Exercise 1.25** Prove Theorem 1.61.

*Hint*: prove that

$$\frac{1}{\rho^n} \int_{B_\rho(x)} \frac{|u(y) - \tilde{u}(x) - \langle \varphi(x), y - x\rangle|}{|x - y|}\, dy$$

$$\le \int_0^1 \left( \frac{1}{t^n \rho^n} \int_{B_{t\rho}(x)} |\varphi(y) - \varphi(x)|\, dy + \frac{1}{t^n \rho^n} |D^s u|(B_{t\rho}(x)) \right) dt,$$

and use the fact that for $\mathcal{L}_n$-a.a. $x \in \Omega$ we have at the same time that $u(x) = \tilde{u}(x)$, $x$ is a Lebesgue point for $\varphi$ and $d|D^s u|/d\mathcal{L}_n = 0$ at $x$

# 2

# SPECIAL FUNCTIONS OF BOUNDED VARIATION

## 2.1 SBV functions. A compactness theorem

**Definition 2.1** *A function* $u \in L^1(\Omega)$ *is a* special function of bounded variation *on* $\Omega$ *if its distributional derivative can be written as*

$$Du = f\,\mathcal{L}_n + g\,\mathcal{H}^{n-1}\llcorner K\,,$$

*where* $f \in L^1(\Omega;\mathbf{R}^n)$, $K$ *is a set of* $\sigma$*-finite Hausdorff measure, and* $g$ *belongs to* $L^1(\Omega,\mathcal{H}^{n-1}\llcorner K;\mathbf{R}^n)$. *The space of special functions of bounded variation is denoted by* $SBV(\Omega)$.

**Remark 2.2** In view of Theorems 1.62 and 1.63 we have the following equivalent definitions:

(i) $u \in SBV(\Omega)$;
(ii) $u \in BV(\Omega)$ and $D^c u = 0$;
(iii) $u \in BV(\Omega)$ and

$$Du = \nabla\,\mathcal{L}_n + (u^+ - u^-)\nu_u\mathcal{H}^{n-1}\llcorner S(u)\,;$$

i.e., $f = \nabla u$ and $g\chi_K = (u^+ - u^-)\nu_u\chi_{S(u)}$.

**Theorem 2.3. (SBV Compactness Theorem)** *Let* $(u_k) \subset SBV(\Omega)$ *be a sequence of special functions of bounded variation in* $\Omega$, *and assume that*

(i) *the sequence* $(u_k)$ *is uniformly bounded in the* $BV$ *norm (i.e., it is relatively compact with respect to the weak topology of* $BV(\Omega)$*);*

(ii) *the approximate gradients* $(\nabla u_k)$ *are equi-integrable (i.e., they are relatively compact with respect to the weak topology of* $L^1(\Omega,\mathbf{R}^n)$*);*

(iii) *there exists a function* $\psi : [0,\infty) \to [0,\infty]$ *such that* $\psi(t)/t \to +\infty$ *as* $t \to 0$, *and*

$$\sup_k \int_{S(u_k)} \psi(|u_k^+ - u_k^-|)\,d\mathcal{H}^{n-1} < \infty \qquad \forall k\,. \tag{2.1}$$

*Then we may extract a subsequence (not relabelled)* $(u_k)$ *which converges in* $L^1(\Omega)$ *to some* $u \in SBV(\Omega)$. *Moreover, the Lebesgue part and the jump part of the derivatives converge separately; i.e.,* $D^a u_k \to D^a u$ *and* $D^j u_k \to D^j u$ *weakly in the sense of measures.*

To prove the theorem above we follow the proof by Alberti and Mantegazza [AM]. We define the auxiliary set of functions

$$X(\psi) = \{\phi \in C^1(\mathbf{R}) :\ \|\phi'\|_\infty < +\infty,\ |\phi(s) - \phi(t)| \le \psi(|s-t|)\ \forall\, t,s \in \mathbf{R}\}\,.$$

**Remark 2.4** We can take in particular $\psi = 1$. In this case, up to translations by constants, the set $X(1)$ coincides with $\{\phi \in C^1(\mathbf{R}) : \|\phi'\|_\infty < +\infty, \|\phi\|_\infty \le \frac{1}{2}\}$.

Note that if $u \in BV(\Omega)$ and $\sigma \in \mathcal{M}_+(\Omega)$ is non-trivial then

$$\sup_{\phi \in X(1)} \int_\Omega |\phi'(\tilde{u})| \, d\sigma = +\infty .$$

In fact, with fixed $M > 0$, take

$$\phi_1(z) = \frac{1}{2}\sin(2Mz), \qquad \phi_2(z) = \frac{1}{2}\sin\Big(2Mz + \frac{\pi}{4}\Big),$$

so that $\phi_1, \phi_2 \in X(1)$, and set

$$A = \Big\{x \in \Omega : |\phi_1'(\tilde{u})| > \frac{M}{\sqrt{2}}\Big\}, \qquad B = \Big\{x \in \Omega : |\phi_2'(\tilde{u})| \ge \frac{M}{\sqrt{2}}\Big\} = \Omega \setminus A .$$

We then get

$$\begin{aligned} \frac{M}{\sqrt{2}}\sigma(\Omega) &= \frac{M}{\sqrt{2}}(\sigma(A) + \sigma(B)) \le \int_A |\phi_1'(\tilde{u})| \, d\sigma + \int_B |\phi_2'(\tilde{u})| \, d\sigma \\ &\le \int_\Omega |\phi_1'(\tilde{u})| \, d\sigma + \int_\Omega |\phi_2'(\tilde{u})| \, d\sigma \le 2 \sup_{\phi \in X(1)} \int_\Omega |\phi'(\tilde{u})| \, d\sigma . \end{aligned}$$

By the arbitrariness of $M$, this supremum is $+\infty$, as required.

From the chain-rule formula we immediately get the following inequality:

$$\sup_{\phi \in X(f)} \big|D(\phi(u)) - \phi'(\tilde{u})\,(D^a u + D^c u)\big|(\Omega) \le \int_{S(u)} \psi(|u^+ - u^-|) \, d\mathcal{H}^{n-1} . \quad (2.2)$$

This inequality in a sense characterizes the measure $D^a u + D^c u$, as precised below.

**Lemma 2.5** *Let $\psi$ and $X(\psi)$ be as above. Let $u \in BV(\Omega)$, and let $\lambda$ be an $\mathbf{R}^n$-valued measure on $\Omega$ such that $|\lambda|(S(u)) = 0$ and*

$$\sup_{\phi \in X(\psi)} \big|D(\phi(u)) - \phi'(\tilde{u})\,\lambda\big|(\Omega) < +\infty . \quad (2.3)$$

*Then $\lambda = D^a u + D^c u$.*

**Proof** Set $\mu = D^a u + D^c u - \lambda$. We have to prove that $\mu = 0$. Since $|D^a u|$, $|D^c u|$ and $|\lambda|$ do not charge the set $S(u)$, we obtain that $\mu$ and $\mathcal{H}^{n-1} \llcorner S(u)$ are mutually singular; hence, from (2.3) we get

$$\sup_{\phi \in X(\psi)} \int_\Omega |\phi'(\tilde{u})| \, d|\mu| = \sup_{\phi \in X(\psi)} |\phi'(\tilde{u})\,\mu|(\Omega)$$

$$\leq \sup_{\phi \in X(\psi)} \left|\phi'(\tilde{u})\,\mu + (\phi(u^+) - \phi(u^-))\,\mathcal{H}^{n-1} \llcorner S(u)\right|(\Omega) < +\infty\,.$$

It now can be easily checked that the first supremum is finite if and only if $|\mu| = 0$, since we have no bound on $\|\phi'\|_\infty$ if $\phi \in X(\psi)$ (see Remark 2.4 above for the case $\psi = 1$). The proof is concluded. □

We can prove Theorem 2.3 as a consequence of Lemma 2.5.

**Proof of Theorem 2.3.** It is not a restriction to suppose that $u_j \to u$ in $L^1(\Omega)$ and a.e., $Du_j \rightharpoonup Du$ weakly in the sense of measures, and $\nabla u_j \rightharpoonup g$ weakly in $L^1(\Omega; \mathbf{R}^n)$. We remark that it is enough to prove that $D^a u + D^c u = g$.

Take $\phi \in X(\psi)$. Then, taking into account inequality (2.2) and the fact that $u_j \in SBV(\Omega)$ for every $j$, we get

$$C \geq \int_{S(u_j)} \psi(|u_j^+ - u_j^-|)\, d\mathcal{H}^{n-1} \geq \left|D(\phi(u_j)) - \phi'(u_j)\,\nabla u_j\, \mathcal{L}_n\right|(\Omega)\,, \qquad (2.4)$$

where $C = \sup_k \int_{S(u_k)} \psi(|u_k^+ - u_k^-|)\, d\mathcal{H}^{n-1}$. The functions $\phi(u_j)$ converge to $\phi(u)$ in the weak topology of $BV(\Omega)$, and then the measures $D(\phi(u_j))$ converge to $D(\phi(u))$ weakly as measures. Since $\phi'$ is bounded and continuous, the functions $\phi'(u_j)$ are uniformly bounded and converge to $\phi'(u)$ a.e. Hence, $\phi'(u_j)\,\nabla u_j$ converge to $\phi'(u)\,g$ weakly as measures, and then

$$\left(D(\phi(u_j)) - \phi'(u_j)\,\nabla u_j \mathcal{L}_n\right) \rightharpoonup \left(D(\phi(u)) - \phi'(u)\, g \mathcal{L}_n\right)$$

weakly as measures. Now, (2.4) yields

$$C \geq \liminf_j \left|D(\phi(u_j)) - \phi'(u_j)\,\nabla u_j \mathcal{L}_n\right|(\Omega) \geq \left|D(\phi(u)) - \phi'(u)\, g\right|(\Omega)\,.$$

Eventually, if we take the supremum over all $\phi \in X(f)$, and apply Lemma 2.5, we get $D^a u + D^c u = g$. □

## 2.2 General lower semicontinuity conditions in one dimension

Lower semicontinuity conditions for general functionals defined on $SBV$ take a complex form, due to the possible interaction of the Lebesgue part and the jump part. In this section we state and prove a lower semicontinuity theorem in dimension one which pinpoints some of the difficulties related to a general formulation. Subsequently, we show how the conditions for lower semicontinuity can be simplified if the functionals are independent on horizontal or vertical translations.

**Theorem 2.6** *Let $f : \mathbf{R} \times \mathbf{R} \times \mathbf{R}$ be a Borel function such that $f(t, \cdot, \cdot)$ is lower semicontinuous for a.a. $t \in \mathbf{R}$ and $f(t, u, \cdot)$ is convex for a.a. $t \in \mathbf{R}$ and for all $u \in \mathbf{R}$. Suppose, moreover, that a convex $\phi : \mathbf{R} \to [0, +\infty)$ exists with $\lim_{|t| \to +\infty} \frac{\phi(t)}{|t|} = +\infty$ and $f(x, u, z) \geq \phi(z)$ for all $(t, u, z) \in \mathbf{R}^3$. Let $\vartheta : \mathbf{R}^3 \to$*

$[0,+\infty)$ *satisfy* $\inf\vartheta > 0$, *and let* $I \subset \mathbf{R}$ *be a bounded open interval. Then the functional*

$$F(u) = \begin{cases} \displaystyle\int_I f(t,u,u')\,dt + \sum_{t\in S(u)} \vartheta(t,u(t-),u(t+)) & \text{if } u \in SBV(I) \\ +\infty & \text{if } u \in BV(I)\setminus SBV(I) \end{cases}$$

*is lower semicontinuous with respect to the weak convergence in* $BV(I)$ *if and only if the following condition holds:*

(I) *for every* $t \in I$, $a,b \in \mathbf{R}$ *with* $a \neq b$, $N \in \mathbf{N}$, *for every* $t_j^0,\ldots,t_j^N \in \mathbf{R}$ *with* $t_j^{i-1} < t_j^i$ $(i=1,\ldots,N)$ *and for every* $a_j^i$, $b_j^i \in \mathbf{R}$ $(i=0,\ldots,N)$ *equibounded in* $\mathbf{R}$, *such that* $t_j^i \to t$, $a_j^0 \to a$, $b_j^N \to b$, *we have*

$$\vartheta(t,a,b) \le \liminf_j \left(\sum_{i=0}^N \vartheta(t_j^i,a_j^i,b_j^i) + \sum_{i=1}^N d(t_j^{i-1},b_j^{i-1};t_j^i,a_j^i)\right), \tag{2.5}$$

*where*

$$d(s,x;t,y) = \inf\left\{\int_s^t f(\tau,u,u')\,d\tau :\ u \in W^{1,1}(s,t),\ u(s)=x,\ u(t)=y\right\}, \tag{2.6}$$

*for all* $s,t,x,y \in \mathbf{R}$.

**Remark 2.7** (a) condition (I) does not take into account the value of $\vartheta$ on the "diagonal" $D := \{(t,a,a) :\ t \in I,\ a \in \mathbf{R}\}$;

(b) if we take $N=0$ then condition (I) implies that $\vartheta$ is lower semicontinuous on $(I\times\mathbf{R}\times\mathbf{R})\setminus D$;

(c) if $\vartheta$ is independent from the first variable and we have $f(t,u,z) \le g(t) + \psi(z)$ with $g \in L^1(I)$ and $\psi$ convex and finite, then the "distance function" $d$ can be dropped from (2.5). In this case condition (I) turns out to be equivalent to the lower semicontinuity and the subadditivity of $\vartheta$; i.e., we must require in addition to lower semicontinuity also that

$$\vartheta(a,b) \le \vartheta(a,c) + \vartheta(c,b) \tag{2.7}$$

for all $a,b,c \in \mathbf{R}$ (see Exercise 2.1);

(d) in condition (I) we can replace the requirement that $(a_j^i)$, $(b_j^i)$ be equibounded in $\mathbf{R}$ by the stronger condition that $a_j^i \to a^i$ and $b_j^i \to b^i$ as $j \to +\infty$, with $a^i$, $b^i$ finite and $a^i = b^i$.

**Proof** Suppose that $F$ be lower semicontinuous. Let $u = a + (b-a)\chi_{[t,+\infty)}$, and let $u_j$ be defined by

$$u_j(s) = \begin{cases} a_j^0 & \text{if } s < t_j^0 \\ b_j^N & \text{if } s \geq t_j^N \\ v_j^i(t) & \text{if } t_j^{i-1} \leq s < t_j^i,\ i = 1, \dots, N, \end{cases}$$

where $v_j^i \in W^{1,1}(t_j^{i-1}, t_j^i)$ is a minimum point for the problem defining the quantity $d(t_j^{i-1}, b_j^{i-1}; t_j^i, a_j^i)$. Since $u_j \rightharpoonup u$ weakly in $BV(I)$ we obtain condition (I) from the inequality $F(u) \leq \liminf_j F(u_j)$.

Conversely, suppose that (I) holds, that $u_j \rightharpoonup u$ weakly in $BV(I)$ and $\lim_j F(u_j) < +\infty$. Since $\inf \vartheta > 0$ we can suppose that $\#(S(u_j)) < +\infty$ and that it is independent of $j$. If $\#(S(u_j)) = 0$ then the sequence is weakly converging in $W^{1,1}(I)$ and the lower semicontinuity follows by classical theorems on Sobolev spaces. We can then suppose that $S(u_j) = \{t_j^0, \dots, t_j^N\}$, with $t_j^{i-1} < t_j^i$, and that $t_j^i \to t^i$ as $j \to +\infty$. Let $S = \{t^0, \dots, t^N\}$, and for each $\delta > 0$ let $S_\delta = \{t \in I : \inf_i |t - t^i| \leq \delta\}$. Again $u_j \rightharpoonup u$ weakly in $W^{1,1}(I \setminus S_\delta)$ so that

$$\int_{I \setminus S_\delta} f(\tau, u, u') d\tau \leq \liminf_j \int_{I \setminus S_\delta} f(\tau, u_j, u_j') d\tau.$$

If $S \cap I = \emptyset$ then the proof is completed. Suppose otherwise; let $t \in S \cap I$, and let $i = i_0, i_0 + 1, \dots, i_0 + M$ be the indices such that $t = t^i$. Since $u_j \to u$ uniformly on $L^\infty(I \setminus S_\delta)$ for all $\delta > 0$ and $u_j$ are equi-uniformly continuous on each interval $(t_j^{i-1}, t_j^i)$, we have that $u_j(t_j^{i_0}-) \to u(t-)$ and $u_j(t_j^{i_0+M}+) \to u(t+)$. From condition (I) we obtain immediately that for $j$ large enough

$$\vartheta(t, u(t-), u(t+)) \leq \liminf_j \sum_{i=i_0}^{i_0+M} \vartheta(t_j^i, u(t_j^i-), u(t_j^i+)) + \int_{t-\delta}^{t+\delta} f(\tau, u_j(\tau), u_j'(\tau))\, d\tau.$$

Note that, since $u_j$ are equi-uniformly continuous on each interval $(t_j^{i-1}, t_j^i)$ we have also that $u_j(t_j^{i_0+k-1}+) - u_j(t_j^{i_0+k}-) \to 0$ as $j \to +\infty$ for all $k = 1, \dots, N$.

Summing in $t \in S \cap I$, and using the inequality proven above, we get, since $S(u) \subset S \cap I$, that

$$\int_{I \setminus S_\delta} f(\tau, u, u') d\tau + \sum_{t \in S(u)} \vartheta(t, u(t-), u(t+)) \leq \lim_j F(u_j).$$

The thesis follows letting $\delta \to 0$. □

**Corollary 2.8** *If we extend $F$ as defined above to $+\infty$ on $L^1(I) \setminus BV(I)$ then $F$ is lower semicontinuous with respect to the $L^1(I)$ convergence along sequences equibounded in $L^\infty(I)$.*

**Proof** It suffices to apply Theorem 2.3. □

**Example 2.9** We show with an example that the interaction between the Lebesgue and the jump parts in the general semicontinuity theorem above cannot be neglected.

Take $f(z) = z^2$, and let $\vartheta : \mathbf{R}^3 \to [0, +\infty)$ be defined by

$$\vartheta(t,v,w) = \begin{cases} 1 & \text{if } t=0,\ v=3,\ w=2 \\ & \text{or } t=0,\ v=2,\ w=0 \\ & \text{or } t=4^{-2n-1},\ v=3,\ w=2-2^{-2n-1} \\ & \text{or } t=4^{-2n},\ v=2-2^{-2n},\ w=0 \\ & \\ 3 & \text{otherwise.} \end{cases}$$

The functional

$$F(u) = \int_{-1}^{1} |u'|^2\,dt + \sum_{t\in S(u)} \vartheta(t, u(t-), u(t+))$$

is lower semicontinuous on $SBV(-1,1)$ with respect to the weak $BV$ convergence even though

(i) $\vartheta(0,\cdot,\cdot)$ is not subadditive on $\mathbf{R}^2$. In fact, $\vartheta(0,3,0) = 3 > 2 = \vartheta(0,3,2) + \vartheta(0,2,0)$;

(ii) we have

$$\vartheta(0,3,0) = 3 > 2 = \lim_n \Big(\vartheta(4^{-2n-1}, 3, 2-2^{-2n-1}) + \vartheta(4^{-2n}, 2-2^{-2n}, 0)\Big)\,.$$

In this case, note that we have

$$d(t,a;s,b) = \min\Big\{\Big|\int_t^s |u'|^2\,d\tau\Big| : u(t)=a,\ u(s)=b\Big\} = \frac{(b-a)^2}{|t-s|},$$

so that

$$\lim_n \Big(\vartheta(4^{-2n-1}, 3, 2-2^{-2n-1}) + \vartheta(4^{-2n}, 2-2^{-2n-1}, 0) \\ + d(4^{-2n-1}, 2-2^{-2n-1}; 4^{-2n}, 2-2^{-2n})\Big) = 3\,.$$

**Theorem 2.10** *Let $\phi : \mathbf{R} \to [0, +\infty]$ and $\theta : \mathbf{R} \to [0, +\infty]$ be lower semicontinuous functions with $\phi$ convex and $\theta$ subadditive; i.e.,*

$$\theta(a+b) \le \theta(a) + \theta(b) \tag{2.8}$$

*for all $a, b \in \mathbf{R}$. Suppose that*

$$\lim_{t\to\pm\infty} \frac{\phi(t)}{|t|} = \lim_{t\to 0} \frac{\theta(t)}{|t|} = +\infty.$$

*Then the functional*

$$F(u)=\begin{cases}\displaystyle\int_I \phi(u')\,dt+\sum_{S(u)}\theta(u^+-u^-) & \text{if } u\in SBV(I)\\ +\infty & \text{if } u\in BV(I)\setminus SBV(I)\end{cases}$$

*is weakly lower semicontinuous.*

**Proof** Let $u_j \rightharpoonup u$ in $BV(I)$ with $\sup_j F(u_j)<+\infty$, and $C=\sup_j\|u_j\|_{BV(I)}$. For every $\varepsilon>0$ let

$$S_j^\varepsilon=\{t\in S(u_j):\ |u_j(t+)-u_j(t-)|<\varepsilon\}$$

and

$$u_j^\varepsilon=u_j-\sum_{t\in S_j^\varepsilon}(u_j(t+)-u_j(t-))\chi_{[t,+\infty)}.$$

Note that $\#(S(u_j)\setminus S_j^\varepsilon)\le\frac{C}{\varepsilon}$, independent of $j$, and

$$\begin{aligned}\|u_j^\varepsilon-u_j\|_\infty&\le\sum_{t\in S_j^\varepsilon}|u_j(t+)-u_j(t-)|\\ &\le\frac{\varepsilon}{\widetilde{\theta}(\varepsilon)}\sum_{t\in S_j^\varepsilon}\theta(u_j(t+)-u_j(t-))\le c\frac{\varepsilon}{\widetilde{\theta}(\varepsilon)},\end{aligned}$$

independent of $j$, where $\widetilde{\theta}(t)=\inf\{\theta(\eta):\ |\eta|\le t\}$. Note that we still have $\lim_{\varepsilon\to0+}\widetilde{\theta}(\varepsilon)/\varepsilon=+\infty$. We can suppose that $u_j^\varepsilon\rightharpoonup u^\varepsilon$ in $BV(I)$. By Theorem 2.3 we have $u^\varepsilon\in SBV(I)$. Clearly, $\|u-u^\varepsilon\|_\infty\le c\varepsilon/\widetilde{\theta}(\varepsilon)$. Note that $u'_\varepsilon=u'$. By the previous theorem we obtain that

$$F(u^\varepsilon)\le\liminf_j F(u_j^\varepsilon)\le\liminf_j F(u_j).$$

An easy use of Fatou's Lemma letting $\varepsilon\to0^+$ yields $F(u)\le\liminf_j F(u_j)$ as desired. □

**Corollary 2.11** *If in addition to the hypotheses of the previous theorem, we have* $\lim_{s\to\pm\infty}\theta(s)=+\infty$ *and* $\theta(s)=0$ *only if* $s=0$, *then the functional* $F$, *extended to* $+\infty$ *on* $L^1(I)\setminus BV(I)$, *is lower semicontinuous with respect to the* $L^1(I)$*-convergence.*

**Proof** If $u_j\to u$ in $L^1(I)$ and $\sup_j F(u_j)<+\infty$, then we have $\sup\{|u_j(t+)-u_j(t-)|:\ t\in S(u_j)\}\le c$, independent of $j$. This implies that $\sup_j\|u_j\|_{BV(I)}<+\infty$, so that the previous theorem can be applied. □

### 2.2.1 *Exercises*

**Exercise 2.1** Give a direct proof that if $\vartheta : \mathbf{R}^2 \to [1, +\infty)$ is lower semicontinuous and subadditive then the functional

$$F(u) = \int_a^b |u'|^2 \, dt + \sum_{t \in S(u)} \vartheta(u(t-), u(t+))$$

is lower semicontinuous on $SBV(I)$ with respect to the weak $BV$ convergence.

**Exercise 2.2** Prove that if $\vartheta(u,v) = \theta(u-v)$, with $\theta$ Lipschitz continuous and subadditive, then $F$ defined above is lower semicontinuous on $SBV(I)$ with respect to the a.e. convergence and to the convergence in measure.

**Exercise 2.3** Prove that if $\theta : \mathbf{R} \to [0, +\infty)$ is subadditive and locally bounded then $\theta(z) \le c(1 + |z|)$. Is this true for $\vartheta : \mathbf{R} \times \mathbf{R} \to [0, +\infty)$ subadditive in the sense of (2.7)?

**Exercise 2.4** Prove that if $\theta : [0, +\infty) \to [0, +\infty)$ is subadditive and we have $\lim_{z \to 0^+} \theta(z)/z = L < +\infty$ then $\theta$ is Lipschitz continuous with constant $L$.

**Exercise 2.5** Show that $|\sin z|$, $\arctan |z|$, $\min\{|z|, 1\}$, $\min\{k + \frac{z^2}{k} : k = 1, 2, \ldots\}$ define subadditive functions.

**Exercise 2.6** Prove that if $\sup \theta \le 2 \inf \theta$ then $\theta$ is subadditive.

**Exercise 2.7** Prove that if $\theta : [0, +\infty) \to [0, +\infty)$ is concave then it is subadditive.

**Exercise 2.8** Prove that if $\theta_1$ and $\theta_2$ are subadditive then also $\theta = \theta_1 \vee \theta_2$ is subadditive.

**Exercise 2.9** Let $\theta_1(z) = \frac{5}{4}|z|$ and let

$$\theta_2(z) = \begin{cases} 1 & \text{if } |z| \le 1 \\ k & \text{if } k - \frac{1}{2} \le |z| \le k, \ k \in \mathbf{N} \\ k + 2(|z| - k) & \text{if } k \le |z| \le k + \frac{1}{2}, \ k \in \mathbf{N}. \end{cases}$$

Prove that $\theta_1$ and $\theta_2$ are subadditive, while $\theta = \theta_1 \wedge \theta_2$ is not.

**Exercise 2.10** Prove that the subspace of $BV(\Omega)$ functions such that $Du = D^c u$ (that is, with $\nabla u = 0$ and $S(u) = \emptyset$) are dense in $L^2(\Omega)$ with respect to the strong convergence.

*Hint*: by approximation, it suffices to show that functions of the form $\chi_Q$, where $Q$ is a cube can be approximated by $BV(\Omega)$ functions such that $Du = D^c u$. By reducing to the 1-dimensional case it it sufficient to construct an approximation of the Heaviside function. For this purpose, take $u$ the Cantor-Vitali function on $[0, 1]$, and define $u_k(x) = u(kx)$ on $[0, 1/k]$, extended to 0 on $(-\infty, 0]$, and to 1 on $[1/k, +\infty)$. The sequence $(u_k)$ provides the desired approximation.

## 2.3 A lower semicontinuity theorem in higher dimensions

The lower semicontinuity theorems of the previous section can be generalized in many ways to the higher-dimensional case. We only prove a simple result for isotropic functionals, which will be useful in the sequel.

**Theorem 2.12** *Let $\phi : \mathbf{R} \times \mathbf{R} \to [0, +\infty)$ be a lower semicontinuous symmetric subadditive function; i.e., $\phi$ is lower semicontinuous and*

$$\phi(z, w) = \phi(w, z) \leq \phi(w, y) + \phi(y, z) \text{ for all } w, y, z \in \mathbf{R},$$

*and let*

$$\Phi(u) = \int_{S(u)} \phi(u^+, u^-)\, d\mathcal{H}^{n-1}$$

*be defined for $u \in SBV(\Omega)$. Then we have*

$$\Phi(u) \leq \liminf_j \Phi(u_j)$$

*whenever $(u_k)$ and $u$ satisfy the thesis of Theorem 2.3.*

**Proof** We define

$$\Phi(u, A) = \int_{S(u) \cap A} \phi(u^+, u^-)\, d\mathcal{H}^{n-1}$$

for all $A \subset \Omega$ and $u \in SBV(\Omega)$. Let $(u_k)$ and $u$ satisfy the thesis of Theorem 2.3. We use the notation of Section 1.8.1. With fixed $\xi \in S^{n-1}$ and $y \in \Pi_\xi$, we have

$$\liminf_k \Phi(u_k, A) \geq \liminf_k \int_{\Pi_\xi} \Phi^1(u_{k\xi,y}, A_{\xi,y})\, d\mathcal{H}^{n-1},$$

where

$$\Phi^1(v, I) = \int_I \phi(v^+, v^-)\, dt$$

is defined on $SBV(I)$. Using Fatou's Lemma, the lower semicontinuity of $\Phi^1$, which can be easily obtained from the previous section, and the convergence of $u_{k\xi,y}$ to $u_{\xi,y}$ for a.a. $y$, we then get

$$\begin{aligned}\liminf_k \Phi(u_k, A) &\geq \int_{\Pi_\xi} \Phi^1(u_{\xi,y}, A_{\xi,y})\, d\mathcal{H}^{n-1} \\ &= \int_{S(u)\cap A} \phi(u^+, u^-) |\langle \xi, \nu_u \rangle|\, d\mathcal{H}^{n-1}.\end{aligned}$$

The thesis can be easily obtained by using Proposition 1.16 with

$$\mu(A) = \inf\Big\{\liminf_j \Phi(w_j, A) : w_j \rightharpoonup u, (w_j) \text{ satisfies the thesis of Theorem 2.3}\Big\},$$

$\lambda = \phi(u^+, u^-)\mathcal{H}^{n-1} \llcorner S(u)$ and $\psi_i = |\langle \xi_i, \nu_u \rangle|$, where $(\xi_i)$ is a dense sequence in $S^{n-1}$. □

**Remark 2.13** We can take for example

$$\Phi(u) = \int_{S(u)} \sqrt{|u^+ - u^-|}\, d\mathcal{H}^{n-1},$$

or

$$\Phi(u) = \mathcal{H}^{n-1}(S(u)).$$

**Remark 2.14** Theorem 2.12 can be generalized to non-isotropic functionals of the form

$$\Phi(u) = \int_{S(u)} \phi(u^+, u^-)\, \varphi(\nu_u)\, d\mathcal{H}^{n-1},$$

where $\varphi$ is even, convex, positive and positively homogeneous of degree 1. For example, we can deal with

$$\Phi(u) = \int_{S(u)} |\langle \nu_u, \xi \rangle|\, d\mathcal{H}^{n-1},$$

where $\xi \in \mathbf{R}^n$ is a fixed vector. Note, however, that the lower semicontinuity of this special functional can be obtained directly following the proof of Theorem 2.12.

**Corollary 2.15** *Let $\psi$ be a function satisfying* (iii) *of Theorem* 2.3. *If we have $\phi(z, w) \geq \psi(|z - w|)$ for all $z, w \in \mathbf{R}^n$, and $f : \mathbf{R}^n \to [0, +\infty)$ is a convex function satisfying*

$$\lim_{|z| \to +\infty} \frac{f(z)}{|z|} = +\infty,$$

*then the functional*

$$F(u) = \begin{cases} \displaystyle\int_\Omega f(\nabla u)\, dx + \Phi(u) & \text{if } u \in SBV(\Omega) \\ +\infty & \text{otherwise} \end{cases}$$

*is lower semicontinuous on $BV(\Omega)$ with respect to the weak $BV$ convergence. If in addition $\phi(z, w) \geq c|z - w|$ for all $z, w \in \mathbf{R}^n$ then $F$ is lower semicontinuous on $L^1(\Omega)$.*

## 2.4 GSBV functions. The Mumford-Shah functional

### 2.4.1 *GSBV functions*

The hypotheses of the $SBV$ compactness theorem are in general not fulfilled by sequences lying in the sub-levels of a great variety of functionals. As a consequence, it is necessary to extend the definition of these functionals to $L^1$ functions which are not of bounded variation. The following space will be a natural framework.

**Definition 2.16** *A function $u \in L^1(\Omega)$ is a* generalized function of bounded variation *if for each $T > 0$ the* truncated function $u_T = (-T) \vee (T \wedge u)$ *belongs to $SBV(\Omega)$. The space of these functions will be denoted by $GSBV(\Omega)$.*

*If $u \in GSBV(\Omega)$ then the approximate gradient $\nabla u$ is defined a.e. on $\Omega$. Note that $\nabla u_T = \nabla u$ a.e. on $\{u = u_T\}$ and $\nabla u_T = 0$ a.e. on $\{u \neq u_T\} = \{|u| > T\}$. Moreover, $S(u) = \bigcup S(u_T)$.*

**Remark 2.17** The functions

$$u_1 = \sum_{k=1}^{\infty} \frac{1}{k} \chi_{[1-(1/k),1)}, \qquad u_2 = \sum_{k=1}^{\infty} \chi_{[1-(1/k^2),1)}, \qquad u_3 = \sum_{k=1}^{\infty} k\, \chi_{[1-(1/k^3),1)}$$

belong to $GSBV(0,1) \setminus BV(0,1)$. Clearly $\#(S(u_i)) + \infty$; moreover, we have $\sup |u_3^+ - u_3^-| = +\infty$.

Let $n = 2$. The function $u = \sum_{k=1}^{\infty} k^2 \chi_{B_{1/k^2}(0)}$ belongs to $GSBV(B_1(0)) \setminus BV(B_1(0))$. In this case $\mathcal{H}^1(S(u)) < +\infty$.

**Remark 2.18** Semicontinuity results for functionals defined on $GSBV(\Omega)$ can be obtained as a corollary to the corresponding results on $SBV(\Omega)$. As an example we can consider a subadditive and increasing $\psi$ satisfying condition (iii) of Theorem 2.3, and a function $f : \mathbf{R}^n \to [0, +\infty]$ satisfying the hypotheses of Corollary 2.15 and $f(0) = 0$. Then the functional

$$F(u) = \begin{cases} \displaystyle\int_\Omega f(\nabla u)\, dx + \int_{S(u)} \psi(|u^+ - u^-|)\, d\mathcal{H}^{n-1} & \text{if } u \in GSBV(\Omega) \\ +\infty & \text{otherwise} \end{cases}$$

is lower semicontinuous on $L^1(\Omega)$. In fact, if $u_k \to u$ in $L^1(I)$ and a.e., and $\sup_k F(u_k) < +\infty$ then the truncations $u_{kT}$ converge to $u_T$ and $\sup_k F(u_{kT}) < +\infty$. By Theorem 2.3 and Corollary 2.15 this shows that $u_T \in SBV(\Omega)$ (i.e., $u \in GSBV(\Omega)$) and

$$F(u_T) \leq \liminf_k F(u_k).$$

Letting $T \to +\infty$ we obtain the lower semicontinuity of $F$.

**Remark 2.19** Note that in dimension 1 the functional in the previous remark is finite only in $SBV(\Omega)$. In fact if $F(u) < +\infty$ then $M = \max |u^+ - u^-| < +\infty$, so that

$$\int_{S(u)} |u^+ - u^-|\, d\mathcal{H}^{n-1} \leq c \int_{S(u)} \psi(|u^+ - u^-|)\, d\mathcal{H}^{n-1} < +\infty$$

where $c = \sup\{\psi(t)/t : 0 < t \leq M\}$. This, together with the integrability of $\nabla u$ implies that $u \in BV(I)$.

### 2.4.2 *The Mumford-Shah functional*

**Definition 2.20** *Let $a, b > 0$. Any functional of the form*

$$F(u) = \begin{cases} a\displaystyle\int_\Omega |\nabla u|^2\,dx + b\mathcal{H}^{n-1}(S(u)) & \textit{if } u \in GSBV(\Omega) \\ +\infty & \textit{otherwise} \end{cases}$$

*will be called a* Mumford-Shah functional. *Note that by Remark 2.18 the functional $F$ is lower semicontinuous in $L^1(\Omega)$.*

**Proposition 2.21** *Let $g \in L^\infty(\Omega)$, $p \geq 1$, and let $F$ be the Mumford-Shah functional defined above. Then there exists $u \in SBV(\Omega)$ solution to the problem*

$$\min\Big\{F(u) + \int_\Omega |u-g|^p\,dx : \ u \in SBV(\Omega)\Big\}.$$

**Proof** Note that if $\widetilde{u} = (-\|g\|_\infty) \vee (\|g\|_\infty \wedge u)$ then

$$F(\widetilde{u}) + \int_\Omega |\widetilde{u}-g|^p\,dx \leq F(u) + \int_\Omega |u-g|^p\,dx\,.$$

Hence, we can perform this minimization subject to the condition $\|u\|_\infty \leq \|g\|_\infty$. Let $(u_k)$ be a minimizing sequence. Since it satisfies the hypotheses of Theorem 2.3 we can suppose it converges to $u \in SBV(\Omega)$ weakly in $BV(\Omega)$ and a.e. Moreover, by the lower semicontinuity of the Mumford-Shah functional we have $F(u) \leq \liminf_k F(u_k)$. It suffices to apply Fatou's Lemma to obtain also

$$\int_\Omega |u-g|^p\,dx \leq \liminf_k \int_\Omega |u_k - g|^p\,dx$$

and the thesis. □

**Remark 2.22** Let $u$ be a minimum point as in the previous proposition. The local regularity results proved by De Giorgi, Carriero, and Leaci [DGC] show that $\mathcal{H}^{n-1}(\overline{S}(u)\setminus S(u)) = 0$, while Proposition 5.3(i) of Ambrosio and Tortorelli [AT1] shows that we have $\mathcal{H}^{n-1}(S(u)\cap K) = \mathcal{M}^{n-1}(S(u)\cap K)$ for every compact set $K \subseteq \Omega$, where

$$\mathcal{M}^{n-1}(E) = \lim_{h\to 0+} \frac{|\{x \in \mathbf{R}^n : \ \mathrm{dist}\,(x,E) < h\}|}{2h}$$

is the *Minkowsky content* of the set $E$.

# 3

# EXAMPLES OF APPROXIMATION

In this chapter we treat 1-dimensional approximations of free-discontinuity problems. The general $n$-dimensional results will be obtained in the next chapter following a general approach which allows to reduce to the 1-dimensional case by slicing and approximation techniques.

## 3.1 Γ-convergence: an overview

In this section we introduce the notion of Γ-convergence and state its main properties. In what follows $X = (X, d)$ is a metric space. For a comprehensive introduction to Γ-convergence we refer to Dal Maso [DM] (see also Part II of [BDF]).

**Definition 3.1** *We say that a sequence* $F_j : X \to [-\infty, +\infty]$ Γ-converges *to* $F : X \to [-\infty, +\infty]$ *(as* $j \to +\infty$*) if for all* $u \in X$ *we have*

(i) (liminf inequality) *for every sequence* $(u_j)$ *converging to* $u$

$$F(u) \leq \liminf_j F_j(u_j); \tag{3.1}$$

(ii) (existence of a recovery sequence) *there exists a sequence* $(u_j)$ *converging to* $u$ *such that*

$$F(u) \geq \limsup_j F_j(u_j), \tag{3.2}$$

*or, equivalently by* (3.1),

$$F(u) = \lim_j F_j(u_j). \tag{3.3}$$

*The function* $F$ *is called the* Γ-limit *of* $(F_j)$ *(with respect to d), and we write* $F = \Gamma\text{-}\lim_j F_j$.

**Remark 3.2** Note that if $(F_j)$ Γ-converges to $F$ then so does every its subsequence.

The reason for the introduction of this notion is explained by the following fundamental theorem.

**Theorem 3.3. (Fundamental Theorem of Γ-convergence)** *Let us suppose that* $F = \Gamma\text{-}\lim_j F_j$, *and let a compact set* $K \subset X$ *exist such that* $\inf_X F_j = \inf_K F_j$ *for all* $j$. *Then*

$$\exists \min_X F = \lim_j \inf_X F_j. \tag{3.4}$$

*Moreover, if* $(u_j)$ *is a converging sequence such that* $\lim_j F_j(u_j) = \lim_j \inf_X F_j$ *then its limit is a minimum point for* $F$.

**Proof** Let $(u_j) \subset K$ satisfy

$$\liminf_j F_j(u_j) = \liminf_j \inf_X F_j .$$

There exists a subsequence $(u_{j_k})$ converging to some $u$, and such that

$$\lim_k F_{j_k}(u_{j_k}) = \liminf_j \inf_X F_j .$$

Using (3.1) and Remark 3.2, we obtain

$$\inf_X F \le F(u) \le \liminf_k F_{j_k}(u_{j_k}) = \liminf_j \inf_X F_j .$$

From (3.2) we have that for all $u \in X$

$$\limsup_j \inf_X F_j \le \limsup_j F_j(u_j) \le F(u);$$

hence,

$$\limsup_j \inf_X F_j \le \inf_X F,$$

and (3.3) is proved. If $u_j \to u$ and $\lim_j F_j(u_j) = \lim_j \inf_X F_j$ we can repeat the proof above and obtain also the last statement of the theorem. □

The definition of Γ-convergence can be given pointwise on $X$. It is convenient also to introduce the notion of Γ-lower and upper limit, as follows.

**Definition 3.4** *Let $F_j : X \to [-\infty, +\infty]$ and $u \in X$. We define*

$$\Gamma\text{-}\liminf_j F_j(u) = \inf\{\liminf_j F_j(u_j) : \ u_j \to u\}; \tag{3.5}$$

$$\Gamma\text{-}\limsup_j F_j(u) = \inf\{\limsup_j F_j(u_j) : \ u_j \to u\}. \tag{3.6}$$

*If $\Gamma\text{-}\liminf_j F_j(u) = \Gamma\text{-}\limsup_j F_j(u)$ then the common value is called the Γ-limit of $(F_j)$ at $u$, and is denoted by $\Gamma\text{-}\lim_j F_j(u)$. Note that this definition is in accord with the previous one, and that $F_j$ Γ-converges to $F$ if and only if $F(u) = \Gamma\text{-}\lim_j F_j(u)$ at all points $u \in X$.*

**Remark 3.5** The following statements are easily derived from the definition of Γ-convergence, and their proof is left as an exercise to the interested reader.

(i) If $F = \Gamma\text{-}\lim_j F_j$ and $G$ is a continuous function then

$$F + G = \Gamma\text{-}\lim_j (F_j + G); \tag{3.7}$$

(ii) if $F_j = F_0$ for all $j \in \mathbf{N}$ then

$$\Gamma\text{-}\lim_j F_j = \overline{F}_0, \tag{3.8}$$

where $\overline{F}_0$ is the *lower semicontinuous envelope* of $F_0$; i.e.,

$$\overline{F}_0(u) = \sup\{G(u) : \ G \text{ is lower semicontinuos and } G \le F_0\} \tag{3.9}$$

$$= \inf\{\liminf_j F_j(u_j) : \ u_j \to u\}. \tag{3.10}$$

(iii) if $F_j \to F_0$ decreasingly with $j$ then $\Gamma\text{-}\lim_j F_j = \overline{F}_0$;

(iv) if $F_j$ increases with $j$ then

$$\Gamma\text{-}\lim_j F_j = \sup_j \overline{F}_j = \lim_j \overline{F}_j; \tag{3.11}$$

in particular if $F_j$ is lower semicontinuous for every $j \in \mathbf{N}$ then

$$\Gamma\text{-}\lim_j F_j = \lim_j F_j; \tag{3.12}$$

(v) the Γ-lower and upper limits define lower semicontinuous functions; moreover

$$\Gamma\text{-}\liminf_j F_j(u) = \Gamma\text{-}\liminf_j \overline{F}_j(u),$$

$$\Gamma\text{-}\limsup_j F_j(u) = \Gamma\text{-}\limsup_j \overline{F}_j(u);$$

(vi) if $(X, d)$ is a topological vector space,and $F = \Gamma\text{-}\lim_j F_j$, with $F_j$ convex for all $j$, then $F$ is convex.

**Remark 3.6** We have

$$\Gamma\text{-}\liminf_j F_j(u) = \sup_{U \ni u,\ U \in \mathcal{U}} \liminf_j \inf_{v \in U} F_j(v),$$

$$\Gamma\text{-}\limsup_j F_j(u) = \sup_{U \ni u,\ U \in \mathcal{U}} \limsup_j \inf_{v \in U} F_j(v),$$

where $\mathcal{U}$ is a basis for the topology of $X$. Note that if $X$ is separable, then we can take $\mathcal{U}$ countable. From this observation we immediately obtain the following compactness result.

**Theorem 3.7** *Let $(X, d)$ be separable, and let $F_j : X \to [-\infty, +\infty]$. Then an increasing sequence of integers $(j_k)$ exists such that the $\Gamma\text{-}\lim_k F_{j_k}$ exists.*

**Proof** Choose $(j_k)$ such that

$$\lim_k \inf_{v \in U} F_{j_k}(v)$$

exists for all $U \in \mathcal{U}$. □

**Remark 3.8** We have $\Gamma\text{-}\lim_j F_j = F$ if and only if for every subsequence $(F_{j_k})$ there exists a further subsequence which Γ-converges to $F$.

We can extend the definition of Γ-convergence to families depending on a real parameter. For example, we can treat Γ-limits of families $(F_\varepsilon)$ as $\varepsilon \to 0^+$.

**Definition 3.9** *We say that a sequence* $F_\varepsilon : X \to [-\infty, +\infty]$ Γ-converges *to* $F : X \to [-\infty, +\infty]$ *as* $\varepsilon \to 0^+$ *if for every choice of positive* $(\varepsilon_j)$ *converging to* 0 *the sequence* $(F_{\varepsilon_j})$ Γ*-converges to* $F$. *Equivalently, we require that for all* $u \in X$ *we have*

(i) (liminf inequality) *for every sequence of positive* $(\varepsilon_j)$ *converging to* 0 *and for every sequence* $(u_j)$ *converging to* $u$

$$F(u) \leq \liminf_j F_{\varepsilon_j}(u_j)\,; \tag{3.13}$$

(ii) (existence of a recovery sequence) *for every* $\eta > 0$ *there exists a family* $(u_\varepsilon)$ *converging to* $u$ *as* $\varepsilon \to 0^+$ *such that*

$$F(u) \geq \limsup_{\varepsilon \to 0^+} F_\varepsilon(u_\varepsilon) - \eta. \tag{3.14}$$

## 3.2 Elliptic approximations

### 3.2.1 *Approximation of the perimeter by elliptic functionals*

The first approximation we address is that of the perimeter. In the general $n$-dimensional setting, since sets of finite perimeter in $\Omega$ may be identified with their characteristic functions as a subset of $BV(\Omega)$, we can define the perimeter functional as follows:

$$P(u) = \begin{cases} |Du|(\Omega) = \mathcal{H}^{n-1}(S(u)) & \text{if } u \in \{0,1\} \text{ a.e.} \\ +\infty & \text{otherwise.} \end{cases} \tag{3.15}$$

The functional $P : L^1(\Omega) \to [0, +\infty]$ is lower semicontinuous with respect to the $L^1(\Omega)$-convergence. In this section we show an approximation of $P$ via elliptic functionals in the 1-dimensional case, where

$$P(u) = \begin{cases} \#(S(u)) & \text{if } u \in \{0,1\} \text{ a.e.} \\ +\infty & \text{otherwise.} \end{cases} \tag{3.16}$$

The $n$-dimensional case will be recovered in Chapter 4.

**Theorem 3.10** *Let* $p > 1$ *and* $p' = \frac{p}{p-1}$, *let* $W : \mathbf{R} \to [0, +\infty)$ *be a continuous function such that* $W(z) = 0$ *if and only if* $z \in \{0, 1\}$, *let* $P_\varepsilon : L^1(\Omega) \to [0, +\infty]$ *be defined by*

$$P_\varepsilon(u) = \begin{cases} \dfrac{1}{\varepsilon p'} \displaystyle\int_\Omega W(u(t))\,dt + \dfrac{\varepsilon^{p-1}}{p} \displaystyle\int_\Omega |u'|^p\,dt & \text{if } u \in W^{1,p}(\Omega) \\ +\infty & \text{otherwise,} \end{cases} \tag{3.17}$$

*and let*

$$c_p = \int_0^1 (W(s))^{1/p'}\,ds\,. \tag{3.18}$$

*Then* $P_\varepsilon$ Γ*-converges to* $c_p P$ *with respect to the* $L^1(\Omega)$*-distance.*

**Proof** For the sake of notation, for all open sets $I$ in $\mathbf{R}$ we set

$$P_\varepsilon(u,I)=\begin{cases}\dfrac{1}{\varepsilon p'}\displaystyle\int_I W(u(t))\,dt+\dfrac{\varepsilon^{p-1}}{p}\int_I |u'|^p\,dt & \text{if } u\in W^{1,p}(I)\\ +\infty & \text{otherwise.}\end{cases}\tag{3.19}$$

Let $u\in L^1(\Omega)$, and let $u_j\to u$ in $L^1(\Omega)$ with $\sup_j P_{\varepsilon_j}(u_j)<+\infty$. We can also suppose that $u_j\to u$ a.e. Since

$$\int_\Omega W(u_j)\,dt\le c\,\varepsilon_j\,,$$

we get that $|\{|W(u_j)|>\eta\}|\to 0$ for all $\eta>0$, so that $u_j\to 0$ or $u_j\to 1$ a.e.; hence, $u\in\{0,1\}$ a.e.

If $S(u)=\emptyset$ there is nothing to prove. Suppose that $t_1,\dots,t_N\in S(u)$. Then we can find $a_i^\pm\in\Omega$, $i=1,\dots,N$, such that $(a_i^-,a_i^+)\subset\Omega$,

$$a_i^-<t_i<a_i^+\le a_{i+1}^-\,,$$

there exist the limits

$$\lim_j u_j(a_i^\pm)=u(a_i^\pm)\in\{0,1\}\,,$$

and $u(a_i^-)\neq u(a_i^+)$. We then have

$$P_{\varepsilon_j}(u_j)\ge\sum_{i=1}^N P_{\varepsilon_j}(u_j,(a_i^-,a_i^+)).$$

Let

$$a_i=\frac{1}{2}(a_i^++a_i^-),\qquad v_j^-(t)=u_j(\varepsilon_j(t-a_i)),\qquad T_j^i=\frac{1}{2\varepsilon_j}(a_i^+-a_i^-)\,.$$

A simple change of variable yields:

$$\int_{(a_i^-,a_i^+)}W(u_j(s))\,ds=\int_{(a_i^-,a_i^+)}W\Big(v_j^i\Big(\frac{s}{\varepsilon_j}-a_i\Big)\Big)\,ds=\int_{(-T_j^i,T_j^i)}W(v_j^i(t))\,dt\,,$$

and

$$\int_{(a_i^-,a_i^+)}|u_j'|^p\,ds=\int_{(a_i^-,a_i^+)}\Big|\frac{1}{\varepsilon_j}v_j^i\Big(\frac{s}{\varepsilon_j}-a_i\Big)\Big|^p\,ds=\varepsilon_j^{1-p}\int_{(-T_j^i,T_j^i)}W(v_j^i(t))\,dt\,.$$

We then have

$$P_{\varepsilon_j}(u_j,(a_i^-,a_i^+))=P_1(v_j^i,(-T_j^i,T_j^i)),$$

so that

$$P_{\varepsilon_j}(u_j,(a_i^-,a_i^+))\ge$$

$$\inf_{T\geq 0}\inf\Big\{P_1(v,(-T,T)):\,v\in W^{1,p}(-T,T),\,v(\pm T)=u_j(a_i^{\pm})\Big\}.$$

Note that by Young's inequality we have

$$\frac{1}{p'}W(v)+\frac{1}{p}|v'|^p\geq W^{1/p'}(v)\,|v'|=|\nabla(\Phi(v))|,$$

where

$$\Phi(z)=\int_0^z W^{1/p'}(s)\,ds\,.$$

This implies that if $v\in W^{1,p}(a,b)$ then

$$\int_{(a,b)}\Big(\frac{1}{p'}W(v)+\frac{1}{p}|v'|^p\Big)\,dt\geq\int_{(a,b)}|\nabla(\Phi(v))|\,dt\geq|\Phi(v(b))-\Phi(v(a))|\,.$$

If we set

$$\begin{aligned}\tilde c=\liminf_{(z,w)\to(0,1)}\ \inf_{T\geq 0}\inf\Big\{&P_1(v,(-T,T))\\ &:v\in W^{1,p}(-T,T),\,v(-T)=z,\ v(T)=w\Big\}\end{aligned}$$

then

$$\tilde c\geq\liminf_{(z,w)\to(0,1)}|\Phi(w)-\Phi(z)|=\Phi(1)=c_p>0.$$

We get

$$\liminf\nolimits_j P_{\varepsilon_j}(u_j)\geq\tilde cN\geq c_pN,$$

and by the arbitrariness of $\{t_1,\dots,t_N\}\subset S(u)$,

$$\liminf\nolimits_j P_{\varepsilon_j}(u_j)\geq\tilde c\#(S(u))\geq c_p\#(S(u))\,.$$

This shows that $\Gamma\text{-}\liminf_{\varepsilon\to 0+}P_\varepsilon\geq c_p\,P$.

We now construct a recovery sequence for the $\Gamma$-limsup. It suffices to consider $u=\chi_E$ with $S(u)$ finite. Since our approximation construction modifies the target function $u$ only in a small neighbourhood of $S(u)$, and is invariant under translations (in $t$) and reflections with respect to $u=1/2$, it is not restrictive to suppose that $u=\chi_{[0,+\infty)}$. With fixed $\eta>0$, choose $T_\eta>0$ and $v_\eta\in W^{1,p}(-T_\eta,T_\eta)$ such that

$$P_1(v_\eta,(-T_\eta,T_\eta))\leq\tilde c+\eta,$$

$$0\leq v_\eta(-T_\eta)\leq\eta,\qquad 1-\eta\leq v_\eta(T_\eta)\leq 1$$

(note that by a truncation argument it is not restrictive to suppose that $0\leq v_\eta\leq 1$). We extend $v_\eta$ to $\mathbf{R}$ setting

$$v_\eta(t) = \begin{cases} 0 \vee (v_\eta(-T_\eta) + t + T_\eta) & \text{if } t < -T_\eta \\ v_\eta(t) & \text{if } t \in [-T_\eta, T_\eta] \\ 1 \wedge (v_\eta(T_\eta) + t - T_\eta) & \text{if } t > T_\eta . \end{cases}$$

Note that

$$\begin{aligned} P_1(v_\eta, \mathbf{R}) &= P_1(v_\eta, (-T_\eta, T_\eta)) \\ &\quad + P_1(v_\eta, \mathbf{R} \setminus [-T_\eta, T_\eta]) \leq \tilde{c} + \eta \Big(1 + \frac{2}{p'} \sup\{W(s) : \ s \in [0,1]\}\Big). \end{aligned}$$

Set

$$u_\varepsilon(t) = v_\eta(\frac{t}{\varepsilon}) = \begin{cases} v_\eta(\frac{t}{\varepsilon}) & \text{if } t \in [-\varepsilon(T_\eta + \eta), \varepsilon(T_\eta + \eta)] \\ u(t) & \text{otherwise;} \end{cases}$$

a simple change of variables as above shows that

$$\limsup_{\varepsilon \to 0+} P_\varepsilon(u_\varepsilon) = P_1(v_\eta, \mathbf{R}) \leq \tilde{c} + c\eta .$$

By the arbitrariness of $\eta > 0$ we have $\Gamma\text{-}\limsup_{\varepsilon \to 0+} P_\varepsilon(u) \leq \tilde{c}$. If $u \in BV(\Omega)$ and $u \in \{0,1\}$ a.e., then, repeating the same argument near each point of $S(u)$, we get that

$$\Gamma\text{-}\limsup_{\varepsilon \to 0+} P_\varepsilon(u) \leq \tilde{c}\,\#(S(u)) .$$

It remains to show that $\tilde{c} = c_p$. Choose $v \in W^{1,1}_{\text{loc}}(\mathbf{R})$ satisfying

$$\begin{cases} v' = W(v)^{1/p} & \text{a.e.} \\ v(-\infty) = 0, \ v(+\infty) = 1 , \end{cases} \tag{3.20}$$

where the values $v(\pm\infty)$ are understood as the existence of the corresponding limits. Note that for such $v$ we have

$$\begin{aligned} \frac{1}{p'} W(v) + \frac{1}{p}|v'|^p &= W(v)^{1/p'}|v'| \\ &= |\nabla(\Phi(v))| = \nabla(\Phi(v)) , \end{aligned}$$

since both $\Phi$ and $v$ are increasing. Hence, we get

$$\begin{aligned} c_p &= \lim_{T \to +\infty} (\Phi(v(T)) - \Phi(v(-T))) \\ &= \lim_{T \to +\infty} \int_{(-T,T)} \Big(\frac{1}{p'} W(v) + \frac{1}{p}|v'|^p\Big)\, dt \geq \tilde{c} . \end{aligned}$$

As the converse inequality was shown above, the proof is concluded. □

**Remark 3.11** From the proof above we get that

$$c_p = \min\Big\{\int_{-\infty}^{+\infty} \big(\frac{1}{p'}W(v) + \frac{1}{p}|v'|^p\big)\,dt : \ v(-\infty) = 0,\ v(+\infty) = 1\Big\},$$

and the minimum is achieved precisely on the solutions of (3.20).

**Remark 3.12** From the proof of Theorem 3.10 we obtain that if $W$ is a non-negative function and $u_j$ is a sequence of $W^{1,p}(a,b)$-functions such that $u_j(t_j^1) \to z_1$ and $u_j(t_j^2) \to z_2$ for two sequences of points $t_j^1, t_j^2 \in (a,b)$, then

$$\liminf_j \int_a^b \Big(\frac{1}{\varepsilon_j p'}W(u_j) + \frac{\varepsilon_j^{p-1}}{p}|u_j|^p\Big)\,dt$$

$$\geq \liminf_j \Big|\int_{t_j^1}^{t_j^2} \Big(\frac{1}{\varepsilon_j p'}W(u_j) + \frac{\varepsilon_j^{p-1}}{p}|u_j'|^p\Big)\,dt\Big| \geq \Big|\int_{z_1}^{z_2} W^{1/p'}(s)\,ds\Big|.$$

**Remark 3.13** The statement of Theorem 3.10 holds true with the same proof if we only suppose that $W : \mathbf{R} \to [0,+\infty]$ is continuous on $[0,1]$ and vanishes only at 0 and 1. For example, we can take $W(z) = +\infty$ if $z \notin [0,1]$. This condition is equivalent to impose the restriction $0 \leq u \leq 1$.

**Remark 3.14** We can consider $W : \mathbf{R} \to [0,+\infty)$ vanishing only at $z_1$ and $z_2$. In this case the $\Gamma$-limit of $P_\varepsilon$ is

$$P_{z_1,z_2}(u) = \begin{cases} c_p^{z_1,z_2}\#(S(u)) & \text{if } u \in \{z_1, z_2\} \text{ a.e.} \\ +\infty & \text{otherwise,} \end{cases}$$

where $c_p^{z_1,z_2} = \Big|\int_{z_1}^{z_2} W^{1/p'}(s)\,ds\Big|$.

### 3.2.2 *Exercises*

**Exercise 3.1** Compute $c_2$ and the solutions to (3.20) in the case $W(s) = s^2(1-s)^2$.

**Exercise 3.2** Compute $c_2$ and the solutions to (3.20) in the case
$$W(s) = \begin{cases} s(1-s) & \text{if } 0 \leq s \leq 1 \\ +\infty & \text{otherwise.} \end{cases}$$

**Exercise 3.3** Prove that in the $n$-dimensional case we have for all $\xi \in S^{n-1}$

$$\Gamma\text{-}\liminf_{\varepsilon\to 0+} P_\varepsilon(u) \geq P^\xi(u),$$

where

$$P^{\xi}(u) = \begin{cases} |\langle Du,\xi\rangle|(\Omega) = \int_{S(u)} |\langle \xi,\nu_u\rangle| d\mathcal{H}^{n-1} & \text{if } u \in \{0,1\} \text{ a.e.} \\ +\infty & \text{otherwise.} \end{cases} \tag{3.21}$$

*Hint*: using the notation of the slicing technique in Section 1.8.1, it suffices to note that, if $u_j \to u$ in $L^1(\Omega)$ then, by Fatou's Lemma,

$$\begin{aligned} \liminf_j P_{\varepsilon_j}(u_j) &\geq \liminf_j \int_{\Pi_\xi} P^{\xi,y}_{\varepsilon_j}((u_j)_{\xi,y})\, d\mathcal{H}^{n-1}(y) \\ &\geq \int_{\Pi_\xi} \liminf_j P^{\xi,y}_{\varepsilon_j}((u_j)_{\xi,y})\, d\mathcal{H}^{n-1}(y) \\ &\geq \int_{\Pi_\xi} P^{\xi,y}(u_{\xi,y})\, d\mathcal{H}^{n-1}(y) = P^{\xi}(u), \end{aligned}$$

where

$$P^{\xi,y}_{\varepsilon}(v) = \begin{cases} \dfrac{1}{\varepsilon p'} \displaystyle\int_{\Omega_{\xi,y}} W(v(t))\, dt + \dfrac{\varepsilon^{p-1}}{p} \displaystyle\int_{\Omega_{\xi,y}} |v'|^p\, dt & \text{if } v \in W^{1,p}(\Omega_{\xi,y}) \\ +\infty & \text{otherwise,} \end{cases} \tag{3.22}$$

and

$$P^{\xi,y}(v) = \begin{cases} \#(S(v)) & \text{if } v \in \{0,1\} \text{ a.e. in } \Omega_{\xi,y} \\ +\infty & \text{otherwise} \end{cases} \tag{3.23}$$

are defined on $L^1(\Omega_{\xi,y})$.

### 3.2.3 *Approximation of the Mumford-Shah functional by elliptic functionals*

We now face the problem of approximating the Mumford-Shah functional, using an extra variable to take care of the surface part.

Let $V : [0,1] \to [0,+\infty)$ be a continuous function vanishing only at the point 1, and let $\psi : [0,1] \to [0,+\infty)$ be an increasing lower semicontinuous function with $\psi(0) = 0$, $\psi(1) = 1$, and $\psi(t) > 0$ if $t \neq 0$. We introduce the functionals $G_\varepsilon : L^1(\Omega) \times L^1(\Omega) \to [0,+\infty]$, defined by

$$G_\varepsilon(u,v) = \begin{cases} \displaystyle\int_\Omega (\psi(v)|\nabla u|^2 + \frac{1}{\varepsilon} V(v) + \varepsilon |\nabla v|^2)\, dt & \text{if } u, v \in H^1(\Omega), \\ & \text{and } 0 \leq v \leq 1 \text{ a.e.} \\ +\infty & \text{otherwise.} \end{cases} \tag{3.24}$$

The functionals above $\Gamma$-converge in a sense to the Mumford-Shah functional. We only deal with the 1-dimensional case, where the convergence theorem can be stated as follows. The $n$-dimensional case will be considered in Chapter 4.

**Theorem 3.15** *Let $V, \psi : [0,1] \to [0,+\infty)$ be as above, let the functionals $G_\varepsilon : L^1(\Omega) \times L^1(\Omega) \to [0,+\infty]$ be defined by*

$$G_\varepsilon(u,v) = \begin{cases} \displaystyle\int_\Omega \Big(\psi(v)|u'|^2 + \frac{1}{\varepsilon}V(v) + \varepsilon|v'|^2\Big)\,dt & \text{if } u,v \in H^1(\Omega), \\ & \text{and } 0 \le v \le 1 \text{ a.e.} \\ +\infty & \text{otherwise,} \end{cases} \tag{3.25}$$

*and let*

$$c_V = \int_0^1 \sqrt{V(s)}\,ds\,,$$

*as in (3.18). Then the functionals $G_\varepsilon$ $\Gamma$-converge as $\varepsilon \to 0^+$ to the functional $G : L^1(\Omega) \times L^1(\Omega) \to [0,+\infty]$, defined by*

$$G(u,v) = \begin{cases} \displaystyle\int_\Omega |u'|^2\,dt + 4\,c_V\#(S(u)) & \text{if } u \in SBV(\Omega), \\ & \text{and } v = 1 \text{ a.e.} \\ +\infty & \text{otherwise.} \end{cases} \tag{3.26}$$

**Proof** For the sake of notation we define $G_\varepsilon(u,v,I)$ and $G(u,v,I)$ if $I \subset \Omega$ as follows:

$$G_\varepsilon(u,v,I) = \begin{cases} \displaystyle\int_I \Big(\psi(v)|u'|^2 + \frac{1}{\varepsilon}V(v) + \varepsilon|v'|^2\Big)\,dt & \text{if } u,v \in H^1(\Omega), \\ & \text{and } 0 \le v \le 1 \text{ a.e.} \\ +\infty & \text{otherwise,} \end{cases} \tag{3.27}$$

$$G(u,v,I) = \begin{cases} \displaystyle\int_I |u'|^2\,dt + 4c_V\,\#(S(u)\cap I) & \text{if } u \in SBV(\Omega), \\ & \text{and } v = 1 \text{ a.e.} \\ +\infty & \text{otherwise.} \end{cases} \tag{3.28}$$

We check the lower semicontinuity inequality for the $\Gamma$-limit. Let $\varepsilon_j \to 0^+$, $u_j \to u$ and $v_j \to v$ in $L^1(\Omega)$. Up to subsequences we can suppose that also $u_j \to u$ and $v_j \to v$ a.e., and that there exists the

$$\lim_j G_{\varepsilon_j}(u_j, v_j) < +\infty\,.$$

It is clear that we must have $v = 1$ a.e., since otherwise $\int_\Omega V(v_j)\,dt$ does not tend to 0, and $G_{\varepsilon_j}(u_j,v_j) \to +\infty$.

We first show that $\#(S(u)) < +\infty$, and

$$4c_V\#(S(u)\cap I) \le \liminf_j G_{\varepsilon_j}(u_j, v_j, I)\,, \tag{3.29}$$

if $I$ is any open subset of $\Omega$. If $S(u) = \emptyset$ there is nothing to prove. Otherwise, choose $\{t_1, \dots, t_N\} \subset S(u)$, and disjoint intervals $I_i = (a_i, b_i) \subset \Omega$ with $t_i \in I_i$. We show that for all $i = 1, \dots, N$

$$\liminf_j G_{\varepsilon_j}(u_j, v_j, I_i) \geq 4c_V \,. \tag{3.30}$$

Let $t_i \in I_i' \subset\subset I_i$, and let $m_i = \liminf_j \inf_{t \in I_i'} \psi(v_j(t))$. If $m_i > 0$ then we have

$$\int_{I_i'} |u_j'|^2 \, dt \leq \frac{1}{m_i} \int_{I_i'} \psi(v_j)|u_j'|^2 \, dt \leq \frac{c}{m_i} \,,$$

so that $u_j \rightharpoonup u$ weakly in $H^1(I_i')$ and $S(u) \cap I_i' = \emptyset$. Hence, we must have $m_i = 0$, and there exists a sequence $(s_j^i) \subset I_i'$ such that $v_j(s_j^i) \to 0$. Moreover, as $v_j \to 1$ a.e. there exist $r_i, r_i' \in I_i$ such that $r_i < s_j^i < r_i'$, and $v_j(r_i') \to 1$, $v_j(r_i) \to 1$. Applying Remark 3.12 twice, we obtain (3.30). Similarly, if $\{t_1, \dots, t_N\} \subset S(u)$ we obtain

$$4c_V N \leq \liminf_j G_{\varepsilon_j}(u_j, v_j, I) \,, \tag{3.31}$$

if $I$ is any open set with $\{t_1, \dots, t_N\} \subset I$, and by the arbitrariness of the choice of $\{t_1, \dots, t_N\} \subset S(u)$ we get (3.29).

Let now $I = (a, b)$ with $I \cap S(u) = \emptyset$. We show that $u \in H^1(I)$, and

$$\int_I |u'|^2 \, dt \leq \liminf_j G_{\varepsilon_j}(u_j, v_j, I) \,. \tag{3.32}$$

Let

$$I_N^k = \Big(a + (b-a)\frac{k-1}{N}, a + (b-a)\frac{k}{N}\Big);$$

up to passing to a subsequence we can suppose that the limits

$$\lim_j \inf_{I_N^k} v_j$$

exist for all $N \in \mathbf{N}$ and $k \in \{1, \dots, N\}$. Let $0 < z < 1$ and $N \in \mathbf{N}$. We consider the set

$$J_N^z = \Big\{k \in \{1, \dots, N\} : \ \lim_j \inf_{t \in I_N^k} v_j(t) \leq z\Big\}.$$

Note that by Remark 3.12

$$\#(J_N^z) \leq c\Big(\int_z^1 \sqrt{V(s)} \, ds\Big)^{-1} ,$$

independent of $N$. We can suppose that $J_N^z = \{k_i : \ i = 1, \dots, L\}$, with $L$ independent of $N$, and that $k_i/N \to t_i \in [a, b]$. Let $S = \{t_1, \dots, t_L\}$. For every $\eta > 0$ we have

$$I_N^k \subset S + [-\eta, \eta]$$

for all $N > 1/\eta$ and $k \in J_N^z$. Hence

$$\begin{aligned}\liminf_j \psi(z)\int_{I\setminus(S+[-\eta,\eta])}|u_j'|^2\,dt &\le \liminf_j \sum_{k\notin J_N^z}\int_{I_N^k}\psi(z)|u_j'|^2\,dt\\ &\le \liminf_j \sum_{k\notin J_N^z}\int_{I_N^k}\psi(v_j)|u_j'|^2\,dt\\ &\le \liminf_j \int_I \psi(v_j)|u_j'|^2\,dt\\ &\le \liminf_j G_{\varepsilon_j}(u_j,v_j,I)\,.\end{aligned}$$

We have that $u \in H^1(I\setminus(S+[-\eta,\eta]))$, and that

$$\psi(z)\int_{I\setminus(S+[-\eta,\eta])}|u_j'|^2\,dt \le \liminf_j G_{\varepsilon_j}(u_j,v_j,I)\,.$$

By the arbitrariness of $\eta > 0$ we obtain that $u \in H^1(I\setminus S)$. Since $I\cap S(u)=\emptyset$ we have that $u\in H^1(I)$; eventually we get (3.32) letting $z \to 1$.

Finally, for all $\eta > 0$ we set $I_\eta^0 = \Omega\setminus(S(u)+[-\eta,\eta])$, $I_\eta^1 = S(u)+(-\eta,\eta)\cap\Omega$. We then have

$$\begin{aligned}\int_{I_\eta}|u'|^2\,dt + 4c_V\,\#(S(u)) &\le \liminf_j G_{\varepsilon_j}(u_j,v_j,I^0,\eta)\\ &\quad+\liminf_j G_{\varepsilon_j}(u_j,v_j,I^1,\eta) \le \liminf_j G_{\varepsilon_j}(u_j,v_j)\,.\end{aligned}$$

Letting $\eta \to 0$ we obtain the lower semicontinuity inequality.

We now turn to the construction of a recovery sequence. It suffices to consider the case $\Omega=(-1,1)$, $u\in SBV(\Omega)$, $u'\in L^2(\Omega)$ and $S(u)=\{0\}$. Choose $\xi_\varepsilon = o(\varepsilon)$, and let $u_\varepsilon \in H^1(\Omega)$ with $u_\varepsilon(t)=u(t)$ if $|t|>\xi_\varepsilon$. With fixed $\eta>0$, let $T>0$ and $v\in H^1(0,T)$ be such that

$$\int_0^T (V(v)+|v'|^2)\,dt \le 2c_V+\eta,$$

$v(0)=0$, $v(T)=1$. We set

$$v_\varepsilon(t)=\begin{cases}0 & \text{if } |t|\le\xi_\varepsilon\\ v\Big(\dfrac{|t|-\xi_\varepsilon}{\varepsilon}\Big) & \text{if } \xi_\varepsilon<|t|<\xi_\varepsilon+\varepsilon T\\ 1 & \text{if } |t|\ge \xi_\varepsilon+\varepsilon T\,.\end{cases}$$

We then get

$$G_\varepsilon(u_\varepsilon,v_\varepsilon)=\int_{-1}^1\Big(\psi(v_\varepsilon)|u'|^2+\frac{1}{\varepsilon}V(v_\varepsilon)+\varepsilon|v_\varepsilon'|^2\Big)\,dt$$

$$\leq \int_{-1}^{1} (|u'|^2 + \frac{1}{\varepsilon}V(v_\varepsilon) + \varepsilon|v_\varepsilon'|^2)\,dt$$
$$\leq \int_{-1}^{1} |u'|^2\,dt + 4c_V + 2\eta + 2V(0)\frac{\xi_\varepsilon}{\varepsilon},$$

so that
$$\limsup_{\varepsilon\to 0+} G_\varepsilon(u_\varepsilon, v_\varepsilon) \leq \int_{-1}^{1} |u'|^2\,dt + 4c_V + 2\eta.$$
By the arbitrariness of $\eta$ we conclude the proof. □

**Exercise 3.4** Let $k_\varepsilon > 0$ and $g \in L^2(\Omega)$. Prove that the problem

$$\min\Big\{G_\varepsilon(u,v) + \int_\Omega |u-g|^2\,dt : u, v \in H^1(\Omega)\Big\}$$

admits a minimizing pair.

*Hint*: $G_\varepsilon$ is coercive with respect to the weak convergence in $H^1(\Omega) \times H^1(\Omega)$ on bounded sets of $L^2(\Omega) \times L^2(\Omega)$.

**Exercise 3.5** Prove that if we substitute $\psi(v)$ by $\psi(v) + k_\varepsilon$ in the definition of $G_\varepsilon$ we obtain the same limit, provided that $k_\varepsilon = o(\varepsilon)$ and $k_\varepsilon \geq 0$.

*Hint*: we only have to prove the existence of recovery sequence, since the lower semicontinuity inequality follows by comparison. In this case take $\xi_\varepsilon = \sqrt{k_\varepsilon \varepsilon}$ and $u_\varepsilon$ affine in $(-\xi_\varepsilon, \xi_\varepsilon)$.

### 3.2.4 *Approximation of free-discontinuity problems by elliptic functionals*

The approach described in the previous section cannot be easily modified to have an approximation of more general free-discontinuity energies. In this section we show, instead, how a large class of functionals can be approximated by a *double limit* procedure.

Let $\varphi, \vartheta : [0, +\infty) \to [0, +\infty)$ be functions satisfying

(i) $\varphi$ is convex and increasing, $\lim_{t\to+\infty} \varphi(t)/t = +\infty$;

(ii) $\vartheta$ is concave, $\lim_{t\to 0+} \vartheta(t)/t = +\infty$.

Let $\Omega$ be a bounded open set of $\mathbf{R}$. The functional $F : L^1(\Omega) \to [0, +\infty]$ defined by

$$F(u) = \begin{cases} \displaystyle\int_\Omega \varphi(|u'|)\,dx + \sum_{S(u)} \vartheta(|u^+ - u^-|) & \text{if } u \in SBV(\Omega) \\ +\infty & \text{otherwise} \end{cases} \tag{3.33}$$

is lower semicontinuous with respect to the $L^1(\Omega)$-convergence by Remark 2.18. We now construct a two-parameter family $G_{j,\varepsilon}$ which approximates $F$.

Let $\vartheta_j : [0, +\infty) \to [0, +\infty)$ be functions of the form

$$\vartheta_j(z) = \min\{A_i^j z + B_i^j : i = 0, \ldots, j\},$$

with $0 = A_0^j < \cdots < A_j^j = j$ and $0 = B_j^j < B_{j-1}^j < \cdots < B_0^j = \max \vartheta_j$, converging increasingly to $\vartheta$, and let $\varphi_j : [0,+\infty) \to [0,+\infty)$ be convex increasing functions with

$$\lim_{t\to+\infty} \frac{\varphi_j(t)}{t} = j,$$

converging increasingly to $\varphi$, and assume that $\varphi(0) = 0$ and $\varphi_j(t) \geq jt - c_j$. We denote $k_j = \max \vartheta_j$. Let $V : [0,1] \to [0,+\infty)$ be a continuous decreasing function vanishing only at the point 1, with $4\int_0^1 \sqrt{V(\tau)}\,d\tau = 1$, and for $s \in [0,1]$ let

$$c_V(s) = 4\int_s^1 \sqrt{V(\tau)}\,d\tau \in [0,1].$$

Let $\psi_j : [0,1] \to [0,1]$ be defined by $\psi_j(0) = 0$ and

$$\psi_j(\xi) = \frac{A_i^j}{j} \qquad \text{if } B_i^j \leq k_j\, c_V(\xi) < B_{i-1}^j\,.$$

Note that for all $z \geq 0$

$$\min\{\psi_j(y)jz + c_V(y) :\ 0 \leq y \leq 1\} = \vartheta_j(z).$$

We consider the functionals $G_{j,\varepsilon} : L^1(\Omega) \times L^1(\Omega) \to [0,+\infty]$ defined by

$$G_{j,\varepsilon}(u,v) = \begin{cases} \displaystyle\int_\Omega \Big(\psi_j(v)\varphi_j(|u'|) + \frac{k_j}{\varepsilon}V(v) + k_j\varepsilon|v'|^2\Big)\,dt & \text{if } u,v \in H^1(\Omega), \\ & \text{and } 0 \leq v \leq 1 \text{ a.e.} \\ +\infty & \text{otherwise.} \end{cases} \tag{3.34}$$

We have the following result.

**Theorem 3.16** *Let the functionals $G_\varepsilon^j$ be defined in* (3.34) *and let*

$$G_j' = \Gamma\text{-}\liminf_{\varepsilon\to0+} G_{j,\varepsilon}, \qquad G_j'' = \Gamma\text{-}\limsup_{\varepsilon\to0+} G_{j,\varepsilon}.$$

*Then we have*

$$G = \Gamma\text{-}\liminf_j G_j' = \Gamma\text{-}\limsup_j G_j'',$$

*where the functional $G : L^1(\Omega) \times L^1(\Omega) \to [0,+\infty]$ is defined by*

$$G(u,v) = \begin{cases} F(u) & \text{if } v = 1 \text{ a.e.} \\ +\infty & \text{otherwise,} \end{cases} \tag{3.35}$$

*with $F$ defined in* (3.33).

**Proof** We use the notation $G_{j,\varepsilon}(u,v,I)$, $G'_j(u,v,I)$, $G''_j(u,v,I)$ and $G(u,v,I)$ for the functionals localized as in (3.27) and (3.28).

Let $j \in \mathbf{N}$ and let $z \in (c_V^{-1}(B^j_{j-1}/k_j), 1)$ be fixed, so that $\psi_j(z) = 1$. Let $\varepsilon_k \to 0$ and let $u_k \to u$ and $v_k \to v$ in $L^1(\Omega)$, and a.e., be such that

$$\lim_k G_{j,\varepsilon_k}(u_k, v_k) = G'_j(u,v) < +\infty.$$

As in the first part of the proof of Theorem 3.15, we deduce that $v = 1$ a.e., and we can find a finite set $S = \{t_i, \dots, t_N\} \subset \Omega$ such that for all $\eta > 0$ we have $v_k > z$ on $\Omega \setminus (S + [-\eta, \eta])$ for $k$ large enough. For fixed $\eta > 0$ we may also suppose that $u_k(t_i \pm \eta) \to u(t_i \pm \eta)$ and $v_k(t_i \pm \eta) \to 1$ for all $i$. Denoting

$$m_k = \inf_{(t_i-\eta, t_i+\eta)} v_k,$$

we then have

$$\begin{aligned}
G_{j,\varepsilon_k}(u_k, v_k, (t_i-\eta, t_i+\eta)) &= \int_{t_i-\eta}^{t_i+\eta} \Big(\psi_j(v_k)\varphi_j(|u_k'|) + \frac{k_j}{\varepsilon_k} V(v_k) + \varepsilon |v_k'|^2\Big)\,dt \\
&\geq \inf_{(t_i-\eta,t_i+\eta)} \psi_j(v_k) \int_{t_i-\eta}^{t_i+\eta} \varphi_j(|u_k'|)\,dt \\
&\quad + 4\int_{m_k}^{\min\{v_k(t_i-\eta), v_k(t_i+\eta))\}} \sqrt{V(s)}ds \\
&\geq \psi_j(m_k)\, 2\eta\, \varphi_j\Big(\frac{|u_k(t_i-\eta) - u_k(t_i+\eta)|}{2\eta}\Big) \\
&\quad + 4\int_{m_k}^{\min\{v_k(t_i-\eta), v_k(t_i+\eta)\}} \sqrt{V(s)}ds, \\
&\geq \psi_j(m_k) j |u_k(t_i-\eta) - u_k(t_i+\eta)| - 2c_j\eta \\
&\quad + 4\int_{m_k}^{\min\{v_k(t_i-\eta), v_k(t_i+\eta)\}} \sqrt{V(s)}ds,
\end{aligned}$$

where we have used Jensen's inequality and $\varphi_j(t) \geq jt - c_j$.

Letting $k \to +\infty$ we have

$$\begin{aligned}
&\liminf_k G_{j,\varepsilon_k}(u_k, v_k, (t_i-\eta, t_i+\eta)) \\
&\geq \min\{\psi_j(y)\, j |u(t_i-\eta) - u(t_i+\eta)| + c_V(y) : 0 \leq y \leq 1\} - 2c_j\eta \\
&= \vartheta_j(|u(t_i-\eta) - u(t_i+\eta)|) - 2c_j\eta,
\end{aligned}$$

and we deduce that $u \in BV(\Omega \setminus (S + [-\eta, \eta]))$ and

$$\begin{aligned}
G'_j(u,1) &\geq \liminf_k G_{j,\varepsilon_k}(u_k, v_k, \Omega \setminus (S + [-\eta,\eta])) \\
&\quad + \sum_{i=1}^N \liminf_k G_{j,\varepsilon_k}(u_k, v_k, (t_i-\eta, t_i+\eta))
\end{aligned}$$

$$
\begin{aligned}
&\geq \liminf_k \int_{\Omega\setminus(S+[-\eta,\eta])} \varphi_j(|u_k'|)\,dt \\
&\quad + \sum_{i=1}^N \vartheta_j(|u(t_i-\eta)-u(t_i+\eta)|) - 2Nc_j\eta \\
&\geq \int_{\Omega\setminus(S+[-\eta,\eta])} \varphi_j(|u'|)\,dt + j|D^s u|(\Omega\setminus(S+[-\eta,\eta])) \\
&\quad + \sum_{i=1}^N \vartheta_j(|u(t_i+)-u(t_i-)|) - 2Nc_j\eta,
\end{aligned}
$$

by using Remark 1.67. By the arbitrariness of $\eta > 0$ we deduce that $u \in BV(\Omega \setminus S)$, and by the finiteness of $S$ that $u \in BV(\Omega)$. Moreover,

$$
G_j'(u,1) \geq \int_\Omega \varphi_j(|u'|)\,dt + j|D^s u|(\Omega\setminus S) + \sum_{i=1}^N \vartheta_j(|u(t_i+)-u(t_i-)|).
$$

As $j|t| \geq \vartheta_j(|t|)$ we obtain

$$
G_j'(u,1) \geq \int_\Omega \varphi_j(|u'|)\,dt + j|D^c u|(\Omega) + \sum_{S(u)} \vartheta_j(|u^+-u^-|). \tag{3.36}
$$

Since the right-hand side converges increasingly to $G(u,1)$ letting $j \to +\infty$, we obtain $\Gamma\text{-}\liminf_j G_j' \geq G$.

To show that $G \geq \Gamma\text{-}\limsup_j G_j''$ we first treat the case of $u$ also satisfying $\#(S(u)) < +\infty$. In this case, we can suppose that $\Omega = (-1,1)$ and $S(u) = \{0\}$. Let $z = |u(0+) - u(0-)|$, and let $y_z \in [0,1]$ be such that

$$
\vartheta_j(z) = \psi_j(y_z)jz + c_V(y_z).
$$

With fixed $\eta > 0$, let $T > 0$ and $v \in H^1(0,T)$ be such that

$$
\int_0^T (V(v) + |v'|^2)\,dt \leq \frac{1}{2}(c_V(y_z) + \eta),
$$

$v(0) = y_z$, $v(T) = 1$. Choose $\xi_\varepsilon = o(\varepsilon)$, and set

$$
u_\varepsilon(t) = \begin{cases}
u(t+\xi_\varepsilon+\varepsilon T) & \text{if } t < -(\xi_\varepsilon+\varepsilon T) \\
u(0-) & \text{if } -\xi_\varepsilon+\varepsilon T < t < -\xi_\varepsilon \\
u(0-) + \dfrac{u(0+)-u(0-)}{2\xi_\varepsilon}(t+\xi_\varepsilon) & \text{if } |t| \leq \xi_\varepsilon + \varepsilon\, \mathrm{T} \\
u(0+) & \text{if } \xi_\varepsilon < t < \xi_\varepsilon + \varepsilon T \\
u(t-\xi_\varepsilon-\varepsilon T) & \text{if } t > \xi_\varepsilon + \varepsilon T
\end{cases}
$$

and

$$v_\varepsilon(t) = \begin{cases} y_z & \text{if } |t| \le \xi_\varepsilon \\ v\Big(\dfrac{|t|-\xi_\varepsilon}{\varepsilon}\Big) & \text{if } \xi_\varepsilon < |t| < \xi_\varepsilon + \varepsilon T \\ 1 & \text{if } |t| \ge \xi_\varepsilon + \varepsilon T\,. \end{cases}$$

We then get, for $\xi_\varepsilon + \varepsilon T < 1$,

$$\begin{aligned} G_{j,\varepsilon}(u_\varepsilon, v_\varepsilon) &= \int_{\Omega\cap\{|t|>\xi_\varepsilon+\varepsilon T\}} \varphi_j(|u_\varepsilon'|)\,dt + \int_{\{|t|<\xi_\varepsilon\}} \psi_j(y_z)\varphi_j(|u_\varepsilon'|)\,dt \\ &\quad + k_j\, 2\int_0^{\varepsilon T} \frac{1}{\varepsilon}\Big(V(v(\frac{t}{\varepsilon})) + |v'(\frac{t}{\varepsilon})|^2\Big)\,dt + \int_{\{|t|<\xi_\varepsilon\}} \frac{1}{\varepsilon}V(v_\varepsilon)\,dt \\ &\le \int_\Omega \varphi(|u'|)\,dt + j\varphi_j(y_z)|u(0+) - u(0-)| \\ &\quad + k_j(c_V(y_z) + \eta) + 2V(0)\frac{\xi_\varepsilon}{\varepsilon}, \end{aligned}$$

so that, since $u_\varepsilon \to u$ and $v_\varepsilon \to 1$ in $L^1(\Omega)$,

$$G_j''(u,1) \le \limsup_{\varepsilon\to 0+} G_{j,\varepsilon}(u_\varepsilon, v_\varepsilon) \le \int_\Omega \varphi(|u'|)\,dt + \vartheta(|u(0+) - u(0-)|) + k_j\eta\,.$$

By the arbitrariness of $\eta$ we obtain that $G_j''(u,1) \le F(u)$. More generally, if $u \in SBV(\Omega)$ then we may find a sequence $(u_k)$ of functions $u_k \in SBV(\Omega)$ with $\#(S(u_k)) < +\infty$ and $\lim_k F(u_k) = F(u)$. By the lower semicontinuity of the $\Gamma$-limsup we then have

$$G_j''(u,1) \le \liminf{}_k G_j''(u_k,1) = \lim{}_k F(u_k) = F(u),$$

and, by the arbitrariness of $j$, $F(u) \ge \Gamma\text{-}\limsup_j G_j''(u,1)$, and hence also $G \ge \Gamma\text{-}\limsup_j G_j''$. The two inequalities yield the thesis. □

**Remark 3.17** It can actually be proven that the $\Gamma$-limit $\Gamma\text{-}\lim_{\varepsilon\to 0+} G_{j,\varepsilon}(u,v)$ exists and is finite only at pairs $(u,1)$ with $u \in BV(\Omega)$, in which case it equals the right-hand side of (3.36) (see [ABS]).

**Remark 3.18** Another way to introduce a non-constant jump energy density by a *single limit* of functionals of the Ambrosio-Tortorelly type is to consider functionals $G_\varepsilon : L^1(\Omega)\times L^1(\Omega) \to [0,+\infty]$ defined by

$$G_\varepsilon(u,v) = \begin{cases} \displaystyle\int_\Omega \Big(\psi(|v|)|u'|^2 + \frac{1}{\varepsilon}V(|u-v|) + \varepsilon|v'|^2\Big)\,dt & \text{if } u,v \in H^1(\Omega), \\ +\infty & \text{otherwise,} \end{cases} \tag{3.37}$$

where now $\psi, V : [0,+\infty) \to [0,+\infty)$ are increasing functions vanishing only at 0 with $V$ continuous and $\psi$ lower semicontinuous. In this case, the $\Gamma$-limit $G$ is finite only on pairs $(u, u)$ with $u \in L^1(\Omega)$ with the property that $\max\{u, z\}$ and $\min\{u, -z\}$ belong to $SBV(\Omega)$ for all $z > 0$, in which case

$$G(u, u) = \int_\Omega \psi(|u|)|u'|^2\, dt + \sum_{S(u)} \vartheta(u^+, u^-),$$

where $\vartheta(a, b) = 2\Big(\int_0^{|a|} \sqrt{V(s)}ds + \int_0^{|b|} \sqrt{V(s)}ds\Big)$ (see [AB]). Note that in particular we can take $\psi(z) = 1$ if $z \neq 0$.

## 3.3 Approximations by high-order perturbations

### 3.3.1 *Surface energies generated by high-order singular perturbation*

In this section and the following one, we face the problem of approximating functionals in SBV, and the Mumford-Shah functional in particular, by functionals depending on only one function variable. It is clear that approximation with local integral functionals depending only on first derivatives is not possible. In fact, from standard convex analysis arguments, the lower semicontinuous envelope with respect to the $L^1(\Omega)$-convergence of functionals of the form

$$F_\varepsilon(u) = \int_\Omega f_\varepsilon(\nabla u)\, dx \qquad u \in W^{1,1}(\Omega)$$

is given by

$$\overline{F}_\varepsilon(u) = \int_\Omega f_\varepsilon^{**}(\nabla u)\, dx \qquad u \in W^{1,1}(\Omega).$$

where $f_\varepsilon^{**}$ denotes the lower semicontinuous and convex envelope of $f^{**}$. Hence, by Remark 3.5(v) and (vi) the $\Gamma$ limit of functionals as above must be convex, and it can be easily checked that functionals on $SBV(\Omega)$ cannot be convex.

The convexity in the gradient variable can be overcome considering higher-order derivatives. Since high derivatives do not appear in the limit, they may be introduced as a singular perturbation. A first result is given by the following theorem.

**Theorem 3.19** *Let $p > 1$, and let $f : [0, +\infty) \to [0, +\infty)$ be a lower semicontinuous increasing function, such that $\alpha, \beta \in \mathbf{R}$ exist with*

$$\alpha = \lim_{t \to 0+} \frac{f(t)}{t}, \qquad \beta = \lim_{t \to +\infty} f(t). \tag{3.38}$$

*Let $I$ be a bounded open subset of $\mathbf{R}$, and let $F_\varepsilon : L^1(I) \to [0, +\infty]$ be defined by*

$$F_\varepsilon(u) = \begin{cases} \dfrac{1}{\varepsilon} \displaystyle\int_I f(\varepsilon|u'|^2)\, dt + \varepsilon^{2p-1} \int_I |u''|^p\, dt & \text{if } u \in W^{2,p}(I) \\ +\infty & \text{otherwise.} \end{cases} \tag{3.39}$$

*Then there exists the* $\Gamma$*-limit* $\Gamma\text{-}\lim_{\varepsilon\to 0^+} F_\varepsilon(u) = F(u)$ *with respect to the* $L^1(I)$*-convergence, where*

$$F(u) = \begin{cases} \alpha \displaystyle\int_I |u'|^2\,dx + m(\beta) \sum_{S(u)} \sqrt{|u^+ - u^-|} & \text{if } u \in SBV(I) \\ +\infty & \text{otherwise,} \end{cases} \tag{3.40}$$

*and*

$$m(s) = \min_{T>0} \min\Big\{ 2sT + \int_{-T}^{T} |\varphi''|^p\,dt : \ \varphi(\pm T) = \pm 1/2, \varphi'(\pm T) = 0 \Big\} \tag{3.41}$$

*for all* $s > 0$.

**Remark 3.20** The choice of the power $\varepsilon^{2p-1}$ follows from the scaling argument leading to the definition of $m(b)$, which will be clear in the proofs (see also Exercise 3.6).

**Remark 3.21** For all $z \in \mathbf{R}$ and $s > 0$ let

$$m(s,z) = \min_{\eta>0} \min\Big\{ 2s\eta + \int_{-\eta}^{\eta} |v''|^p\,dt : \ v \in W^{2,p}(-\eta,\eta), \\ v(\pm\eta) = \pm\frac{z}{2},\ v'(\pm\eta) = 0 \Big\}.$$

If we set

$$c_p = \min\Big\{ \int_{-1}^{1} |\varphi''|^p\,dt : \varphi \in W^{2,p}(-1,1),\ \varphi(\pm 1) = \pm\frac{1}{2},\ \varphi'(\pm 1) = 0 \Big\},$$

then the substitution $v(t) = z\,\varphi(t/\eta)$ gives

$$\begin{aligned} m(s,z) &= \min_{\eta>0} \Big\{ 2\eta s + |z|^p \eta^{1-2p} c_p \Big\} \\ &= s^{(2p-1)/2p} \sqrt{|z|} \Big( 2^{2p-1} c_p (2p-1) \Big)^{1/2p} \Big( 1 + \frac{1}{c_p(2p-1)} \Big) \\ &= s^{(2p-1)/2p} m(1) \sqrt{|z|} = m(s)\sqrt{|z|}. \end{aligned}$$

If $p = 2$ then $c_2$ is easily computed, noticing that the solution $\varphi$ is the third order polynomial satisfying the given boundary conditions. In this case, $m(s) = s^{\frac{3}{4}} \Big( 2\sqrt{3/2} + \sqrt{2/3} \Big)$.

The proof of Theorem 3.19 will be obtained as a consequence of some simpler propositions which deal with lower and upper $\Gamma$-limits separately. Before stating and proving them, we define a "localized version" of our functionals, which highlights their behaviour as set functions, by setting

$$F_\varepsilon(u,I)=\begin{cases}\frac{1}{\varepsilon}\int_I f(\varepsilon|u'|^2)\,dt+\varepsilon^{2p-1}\int_I|u''|^p\,dt & \text{if } u\in W^{2,p}(I)\\ +\infty & \text{otherwise,}\end{cases}\tag{3.42}$$

and

$$F(u,I)=\begin{cases}\alpha\int_I|u'|^2\,dx+m(\beta)\sum_{I\cap S(u)}\sqrt{|u^+-u^-|} & \text{if } u\in SBV(I)\\ +\infty & \text{otherwise,}\end{cases}\tag{3.43}$$

for all $u\in L^1_{\rm loc}(\mathbf{R})$ and $I\subseteq\mathbf{R}$ bounded open set.

**Proposition 3.22** *Let $f(t)=\min\{at,b\}$ and let $u_\varepsilon\in W^{2,p}(I)$. Then there exists $v_\varepsilon\in SBV(I)$ such that*

$$a\int_I|v_\varepsilon'|^2\,dt=\frac{1}{\varepsilon}\int_I f(\varepsilon|u_\varepsilon'|^2)\,dt,\tag{3.44}$$

$$\int_{S(v_\varepsilon)}\sqrt{|v_\varepsilon^+-v_\varepsilon^-|}\,d\#\le\frac{F_\varepsilon(u_\varepsilon)}{m(b)},\tag{3.45}$$

*and* $\|u_\varepsilon-v_\varepsilon\|_{L^1(I)}\le\varepsilon\, c\,(F_\varepsilon(u_\varepsilon))^3$.

**Proof** Set

$$D_\varepsilon=\{t\in I:\ \varepsilon|u_\varepsilon'|^2>b/a\}.$$

Since $D_\varepsilon$ is open, we can write $D_\varepsilon=\bigcup_{k\in\mathbf{N}}I_\varepsilon^k$ as the union of disjoint open intervals $I_\varepsilon^k=(a_\varepsilon^k,b_\varepsilon^k)$. It is not restrictive to suppose that $[a_\varepsilon^k,b_\varepsilon^k]\subset I$ for all $k$. Note that

$$\frac{1}{\varepsilon}|D_\varepsilon|b\le C:=F_\varepsilon(u_\varepsilon).$$

Consider $v_\varepsilon$ defined by

$$v_\varepsilon(t)=\begin{cases}u_\varepsilon(t) & \text{if } t\in I\setminus D_\varepsilon\\ u_\varepsilon(a_\varepsilon^k)+u_\varepsilon'(a_\varepsilon^k)(t-a_\varepsilon^k) & \text{if } t\in I_\varepsilon^k.\end{cases}\tag{3.46}$$

As $|v_\varepsilon'|^2=b/\varepsilon a$ on $D_\varepsilon$, we have

$$a\int_I|v_\varepsilon'|^2\,dt=a\int_{I\setminus D_\varepsilon}|u_\varepsilon'|^2\,dt+\frac{b}{\varepsilon}|D_\varepsilon|=\frac{1}{\varepsilon}\int_I f(\varepsilon|u_\varepsilon'|^2)\,dt\le C;$$

i.e., (3.44).

If $t\in I_\varepsilon^k$ we get, using Hölder's inequality,

$$|v_\varepsilon(t)-u_\varepsilon(t)|=\int_{a_\varepsilon^k}^t|v_\varepsilon'-u_\varepsilon'|\,ds\le\int_{a_\varepsilon^k}^t\int_{a_\varepsilon^k}^s|u_\varepsilon''(\tau)|\,d\tau\,ds$$

$$\leq \left(\int_{I_\varepsilon^k} |u_\varepsilon''|^p\right)^{1/p} \int_{a_\varepsilon^k}^t (s - a_\varepsilon^k)^{(p-1)/p}\, ds$$
$$= \frac{p}{2p-1}\left(\int_{I_\varepsilon^k} |u_\varepsilon''|^p\right)^{1/p} (t - a_\varepsilon^k)^{(2p-1)/p};$$

hence, integrating on $I_\varepsilon^k$,

$$\int_{I_\varepsilon^k} |v_\varepsilon - u_\varepsilon|\, dt \leq c\left(\int_{I_\varepsilon^k} |u_\varepsilon''|^p\right)^{1/p} |I_\varepsilon^k|^{(3p-1)/p}.$$

We eventually obtain

$$\|u_\varepsilon - v_\varepsilon\|_{L^1(I)} = \int_{D_\varepsilon} |v_\varepsilon - u_\varepsilon|\, dt \leq c \sum_{k\in\mathbf{N}} \left(\int_{I_\varepsilon^k} |u_\varepsilon''|^p\right)^{1/p} |I_\varepsilon^k|^{(3p-1)/p}$$
$$\leq c\left(\int_{D_\varepsilon} |u_\varepsilon''|^p\right)^{1/p} \left(\sum_{k\in\mathbf{N}} |I_\varepsilon^k|^{(3p-1)/(p-1)}\right)^{p-1/p}$$
$$\leq c\frac{C^{1/p}}{\varepsilon^{(2p-1)/p}} |D_\varepsilon|^{3p-1/p} \leq c\, C^3\, \varepsilon$$

as required.

The function $v_\varepsilon$ is discontinuous only at the points $b_\varepsilon^k$. Set

$$z_\varepsilon^k = |v_\varepsilon^+(b_\varepsilon^k) - v_\varepsilon^-(b_\varepsilon^k)|, \qquad w_\varepsilon = u_\varepsilon - v_\varepsilon\,.$$

Since $u_\varepsilon'' = w_\varepsilon''$ on $I_\varepsilon^k$, we have

$$\frac{1}{\varepsilon}\int_{I_\varepsilon^k} f(\varepsilon|u_\varepsilon'|^2)\, dt + \varepsilon^{2p-1}\int_{I_\varepsilon^k} |u_\varepsilon''|^p\, dt = \frac{b}{\varepsilon}|I_\varepsilon^k| + \varepsilon^{2p-1}\int_{I_\varepsilon^k} |w_\varepsilon''|^p\, dt$$
$$\geq \min\Big\{\frac{b}{\varepsilon}|I_\varepsilon^k| + \varepsilon^{2p-1}\int_{I_\varepsilon^k} |\varphi''|^p\, dt:\ \varphi \in W^{2,p}(I_\varepsilon^k),\ \varphi(a_\varepsilon^k) = 0,\ \varphi(b_\varepsilon^k) = z_\varepsilon^k,$$
$$\varphi'(a_\varepsilon^k) = \varphi'(b_\varepsilon^k) = 0\Big\}$$
$$\geq \min_{\eta>0}\min\Big\{2\eta b + \int_{-\eta}^{\eta} |\psi''|^p\, dt:\ \psi \in W^{2,p}(-\eta,\eta),\ \psi(\pm\eta) = \pm\frac{z_\varepsilon^k}{2}, \psi'(\pm\eta) = 0\Big\}$$
$$= m(b)\sqrt{z_\varepsilon^k}\,;$$

the last equality being shown in Remark 3.21. Hence,

$$\sum_{k\in\mathbf{N}} \sqrt{|v_\varepsilon^+(b_\varepsilon^k) - v_\varepsilon^-(b_\varepsilon^k)|} \leq \frac{C}{m(b)}\,. \tag{3.47}$$

By (3.44) and (3.47) we get that $v_\varepsilon \in SBV(I)$, $S(v_\varepsilon) = (b_\varepsilon^k)_k$, and (3.45) holds, so that the proof is complete. □

**Proposition 3.23** *Let $f(t) = \min\{at, b\}$. Let $(\varepsilon_j)$ be a sequence of positive numbers converging to 0, let $\sup_j F_{\varepsilon_j}(u_{\varepsilon_j}) < +\infty$, and let $v_{\varepsilon_j}$ be defined as in (3.46). Then there exists a subsequence (not relabeled) of $(\varepsilon_j)$ and a function $u \in SBV(I)$ such that $v_{\varepsilon_j} \to u$ in $L^1(I)$,*

$$v'_{\varepsilon_j} \rightharpoonup u' \quad \textit{weakly in } L^2(I), \tag{3.48}$$

$$\int_{I\cap S(u)} \sqrt{|u^+ - u^-|}\, d\# \le \liminf_j \int_{I\cap S(v_{\varepsilon_j})} \sqrt{|v^+_{\varepsilon_j} - v^-_{\varepsilon_j}|}\, d\#. \tag{3.49}$$

**Proof** By (3.44) and (3.47) we get that $\sup_j \|v_{\varepsilon_j}\|_{BV(I)} < +\infty$ and

$$\sup_j \Big( \int_I |v'_{\varepsilon_j}|^2\, dt + \sum_{I\cap S(v_{\varepsilon_j})} \sqrt{|v^+_{\varepsilon_j} - v^-_{\varepsilon_j}|} \Big) < +\infty.$$

Applying Theorems 2.3 and 2.12, with $\psi(z) = \sqrt{|z|}$, to the sequence $(v_{\varepsilon_j})$ we obtain the existence of $u \in SBV(I)$ such that, up to passing to a subsequence, $v_{\varepsilon_j} \to u$ in $L^1(I)$ and (3.48) and (3.49) hold. □

**Proposition 3.24** *If $f(t) = \min\{at, b\}$ then for all $u \in L^1_{\rm loc}(\mathbf{R})$ we have*

$$F(u, I) \le \Gamma\text{-}\liminf_{\varepsilon\to 0+} F_\varepsilon(u, I),$$

*for all bounded open sets $I \subset \mathbf{R}$, where $F_\varepsilon$ is defined in (3.42) and $F$ in (3.43), with $a, b$ in place of $\alpha, \beta$.*

**Proof** It is not restrictive to suppose that $\Gamma\text{-}\liminf_{\varepsilon\to 0+} F_\varepsilon(u, I) < +\infty$. Let $(\varepsilon_j)$ be a sequence of positive numbers converging to 0 and let $u_{\varepsilon_j} \to u$ be a sequence in $L^1_{\rm loc}(\mathbf{R})$ such that the limit $\lim_j F_{\varepsilon_j}(u_{\varepsilon_j}, I)$ exists and equals the $\Gamma\text{-}\liminf_{\varepsilon\to 0+} F_\varepsilon(u, I)$.

Let $v_{\varepsilon_j}$ be defined by (3.46). Note that $v_{\varepsilon_j} \to u$ in $L^1_{\rm loc}(\mathbf{R})$. With fixed $\delta \in (0, 1)$, by (3.44) and (3.45) (applied with $\delta b$ in place of $b$) we obtain

$$\begin{aligned} F_{\varepsilon_j}(u_{\varepsilon_j}, I) &\ge (1-\delta)\frac{1}{\varepsilon_j}\int_I f(\varepsilon_j |u'_{\varepsilon_j}|^2)\, dt \\ &\quad + \delta \frac{1}{\varepsilon_j}\int_I f(\varepsilon_j |u'_{\varepsilon_j}|^2)\, dt + {\varepsilon_j}^{2p-1}\int_I |u''_{\varepsilon_j}|^p\, dt \\ &\ge (1-\delta) a \int_I |v'_{\varepsilon_j}|^2\, dt + m(\delta b) \sum_{I\cap S(v_{\varepsilon_j})} \sqrt{|v^+_{\varepsilon_j} - v^-_{\varepsilon_j}|}. \end{aligned}$$

By (3.48) and (3.49) we deduce that

$$\liminf_{\varepsilon_j\to 0+} F_{\varepsilon_j}(u_{\varepsilon_j}, I) \ge (1-\delta) a \int_I |u'|^2\, dt + m(\delta b) \sum_{I\cap S(u)} \sqrt{|u^+ - u^-|}.$$

After noting that the open set function $\mu(I) = \Gamma\text{-}\liminf_{\varepsilon\to 0+} F_\varepsilon(u, I)$ is superadditive on disjoint open sets, we can apply Proposition 1.16 with

$$\lambda = \mathcal{L}_1 + \sum_{t\in S(u)} \sqrt{|u^+ - u^-|}\delta_t,$$

and, if $(\delta_i) = \mathbf{Q} \cap (0, 1)$,

$$\psi_i(x) = \begin{cases} (1-\delta_i)a|u'(x)|^2 & \text{a.e. on } I \setminus S(u) \\ m(\delta_i b) = \delta_i^{(2p-1)/2p} m(b) & \text{on } S(u), \end{cases}$$

obtaining the thesis. □

**Proposition 3.25** *Under the hypotheses of Theorem* 3.19 *we have*

$$\Gamma\text{-}\liminf_{\varepsilon\to 0+} F_\varepsilon(u, I) \geq F(u, I)$$

*for all* $u \in L^1_{\rm loc}(\mathbf{R})$, *and for all bounded open sets* $I \subset \mathbf{R}$.

**Proof** Let $(a_i)$ $(b_i)$ be sequences of positive numbers such that $\sup_i a_i = \alpha$, $\sup_i b_i = \beta$, and

$$f_i(t) := \max\{a_i t, b_i\} \leq f(t) \qquad \text{for all } t \geq 0\,.$$

From Proposition 3.24 we have that $\Gamma\text{-}\liminf_{\varepsilon\to 0+} F_\varepsilon(u, I)$ is finite only if $F(u, I)$ is finite, and that

$$\Gamma\text{-}\liminf_{\varepsilon\to 0+} F_\varepsilon(u, I) \geq a_i \int_I |u'|^2\,dt + m(b_i) \sum_{I\cap S(u)} \sqrt{|u^+ - u^-|}\,.$$

The thesis follows as in the proof of Proposition 3.24 taking now

$$\psi_i(x) = \begin{cases} a_i|u'(x)|^2 & \text{a.e. on } I \setminus S(u) \\ m(b_i) & \text{on } S(u) \end{cases}$$

in Proposition 1.16. □

In the sequel $u'(t\pm)$ denote the right-hand side and left-hand side limits of $u'$ at $t$.

**Proposition 3.26** *Let* $u \in SBV(I)$ *satisfy* $\#(S(u)) < +\infty$, $u \in W^{2,p}(I\setminus S(u))$, *and* $u'(t\pm) = 0$ *on* $S(u)$. *Then there exists a family* $(u_\varepsilon)$ *converging to* $u$ *in* $L^1(I)$ *such that* $\limsup_{\varepsilon\to 0+} F_\varepsilon(u_\varepsilon) \leq F(u)$.

**Proof** Since the construction of $u_\varepsilon$ will modify $u$ only in a small neighbourhood of $S(u)$, we can suppose that $I = (-1,1)$ and $S(u) = \{0\}$. Moreover, by a translation argument we can suppose also that $u^+(0) + u^-(0) = 0$. Let $z = u^+(0) - u^-(0)$, and let $\eta$ and $v$ be the minimizing pair in the definition of $m(\beta, z)$. If we set $v_\varepsilon(x) = v(x/\varepsilon)$ then we have $v_\varepsilon(\pm\varepsilon\eta) = \pm z/2$, $v'_\varepsilon(\pm\varepsilon\eta) = 0$, and

$$F_\varepsilon(v_\varepsilon, (-\varepsilon\eta, \varepsilon\eta)) \leq 2\eta\beta + \varepsilon^{2p-1} \int_{-\eta}^{\eta} |v''_\varepsilon|^p \, dt = m(\beta, z).$$

We then define

$$u_\varepsilon(x) = \begin{cases} v_\varepsilon(x) & \text{if } x \in (-\varepsilon\eta, \varepsilon\eta) \\ u(x + \varepsilon\eta) & \text{if } x \leq -\varepsilon\eta \\ u(x - \varepsilon\eta) & \text{if } x \geq \varepsilon\eta, \end{cases}$$

so that $u_\varepsilon \in W^{2,p}(I)$, $u_\varepsilon \to u$ in $L^1(I)$, and

$$F_\varepsilon(u_\varepsilon) \leq \frac{1}{\varepsilon} \int_I f(\varepsilon |u'|^2) \, dt + \varepsilon^{2p-1} \int_I |u''|^p \, dt + m(\beta, z).$$

Note that $f(\varepsilon|u'|^2)/\varepsilon \leq K|u'|^2$ for some constant $K$, and that $f(\varepsilon|u'|^2)/\varepsilon \to \alpha|u'|^2$ a.e. on $I$; hence, after applying Lebesgue's Dominated Convergence Theorem, we obtain

$$\limsup_{\varepsilon \to 0+} F_\varepsilon(u_\varepsilon) \leq \alpha \int_I |u'|^2 \, dt + m(\beta, z) = F(u),$$

and the thesis. □

**Proposition 3.27** *Let $u \in SBV(I)$ satisfy $\#(S(u)) < +\infty$ and $u' \in L^2(I)$. Then there exists a sequence $(u_j)$ in $SBV(I)$ such that $S(u_j) \subseteq S(u)$, $u_j \in W^{2,p}(I \setminus S(u))$, $u'_j(t\pm) = 0$ on $S(u)$, $u_j \to u$ in $L^\infty(I)$, $u'_j \to u'$ in $L^2(I)$ and $u_j(t\pm) \to u(t\pm)$ on $S(u)$.*

**Proof** It is not restrictive to suppose $I = (a,b)$. Let $S(u) = \{x_1, \ldots, x_N\}$, with $x_i < x_{i+1}$, and set $x_0 = a$, $x_{N+1} = b$. Let $(v_j)$ be a sequence of functions in $C^\infty(I \setminus S(u))$ converging strongly to $u$ in $H^1(x_i, x_{i+1})$ for all $i \in \{0, 1, \ldots, N\}$. For all $j \in \mathbf{N}$, and $i \in \{0, 1, \ldots, N\}$, let $u^i_j$ be the solution to the minimum problem

$$\min\Big\{ \int_{x_i}^{x_{i+1}} |v'|^2 \, dt + j \int_{x_i}^{x_{i+1}} |v_j - v|^2 \, dt : \ v \in H^1(x_i, x_{i+1}) \Big\}.$$

Note that $u^i_j$ is also a classical solution of the Euler equation $v'' = j(v - v_j)$ with the Neumann conditions $v'(x_i) = v'(x_{i+1}) = 0$. The function $u_j$ defined by $u_j = u^i_j$ on $(x_i, x_{i+1})$ satisfies the required conditions. Note that $u_j \to u$ in $W^{1,2}(I \setminus S(u))$, and then also in $L^\infty(I)$. In particular $u_j(t\pm) \to u(t\pm)$ on $S(u)$.
□

The following proposition concludes the proof of Theorem 3.19.

**Proposition 3.28** *We have* $\Gamma\text{-}\limsup_{\varepsilon\to 0+} F_\varepsilon(u_\varepsilon) \le F(u)$ *for all* $u \in SBV(I)$.

**Proof** We use the notation $F'' = \Gamma\text{-}\limsup_{\varepsilon\to 0+} F_\varepsilon$, and we suppose without loss of generality that $I = (a,b)$.

If $u \in SBV(I)$ with $\#(S(u)) < +\infty$ and $u' \in L^2(I)$, let $(u_j)$ be given by Proposition 3.27. By Proposition 3.26 we have that $F''(u_j) \le F(u_j)$ for all $j$. Moreover, by Proposition 3.27 $\lim_j F(u_j) = F(u)$. By the lower semicontinuity of $F''$ we then obtain

$$F''(u) \le \liminf_j F''(u_j) \le \liminf_j F(u_j) = F(u). \tag{3.50}$$

Let now $u \in SBV(I)$ satisfy $F(u) < +\infty$ and $\#(S(u)) = +\infty$. If $S(u) = \{x_1, x_2, \ldots\}$, set $z_i = u^+(x_i) - u^-(x_i)$ and

$$u_k = u - \sum_{j=k+1}^{\infty} z_i \chi_{(x_i,b)}.$$

We have $u_k' = u'$, $S(u_k) = \{x_1, \ldots, x_k\}$, $u_k^+(x_i) - u_k^-(x_i) = z_i$ on $S(u_k)$, and $\lim_k F(u_k) = F(u)$. By (3.50) we have $F''(u_k) \le F(u_k)$. Using the lower semicontinuity of $F''$ again, we obtain the required inequality. □

### 3.3.2 *Exercises*

**Exercise 3.6** Let $f$ satisfy the hypotheses of Theorem 3.19 and let $\gamma > 0$, $p > 1$. Let $I$ be a bounded open subset of $\mathbf{R}$, and let $F_\varepsilon^\gamma : L^1(I) \to [0, +\infty]$ be defined by

$$F_\varepsilon(u) = \begin{cases} \dfrac{1}{\varepsilon}\displaystyle\int_I f(\varepsilon|u'|^2)\,dt + \varepsilon^\gamma \int_I |u''|^p\,dt & \text{if } u \in W^{2,p}(I) \\ +\infty & \text{otherwise.} \end{cases} \tag{3.51}$$

Then there exists the $\Gamma$-limit $\Gamma\text{-}\lim_{\varepsilon\to 0+} F_\varepsilon^\gamma(u) = F^\gamma(u)$ with respect to the $L^1(I)$-convergence, where $F^\gamma(u) = 0$ for all $u \in L^1(I)$ if $\gamma > 2p-1$, $F^\gamma = F$ as in Theorem 3.19 if $\gamma = 2p-1$, and

$$F^\gamma(u) = \begin{cases} \alpha \displaystyle\int_I |u'|^2\,dx & \text{if } u \in H^1(I) \\ +\infty & \text{otherwise,} \end{cases} \tag{3.52}$$

if $\gamma < 2p-1$.

*Hint*: use Theorem 3.19 to compare the $\Gamma$-limit with the $\Gamma$-limits of the functionals

$$F_{\eta,\varepsilon}(u) = \eta^{\gamma-2p+1}\Big(\frac{1}{\varepsilon}\int_I \eta^{2p-1-\gamma} f(\varepsilon|u'|^2)\,dt + \varepsilon^{2p-1}\int_I |u''|^p\,dt\Big),$$

where $\eta > 0$ is fixed, first for $u \in W^{2,p}(I)$ and then reasoning by density. Eventually, let $\eta \to 0$.

**Exercise 3.7** Let $g \in L^2(I)$ and $\lambda > 0$ be fixed. Prove that for all $\varepsilon > 0$ there exists a minimum point $u_\varepsilon$ for the problem

$$\min\Big\{\frac{1}{\varepsilon}\int_I f(\varepsilon|v'|^2)\,dt+\varepsilon^{2p-1}\int_I |v''|^p\,dt+\lambda\int_I |v-g|^2\,dt\,:\,v\in W^{2,p}(I)\Big\},\tag{3.53}$$

and for every sequence $(\varepsilon_j)$ of positive numbers converging to 0 there exists a subsequence (not relabeled) such that $u_{\varepsilon_j}$ converges in $L^1(I)$ to a function $u \in SBV(I)$, which minimizes

$$\min\Big\{\alpha\int_I |v'|^2+m(\beta)\int_{S(v)}\sqrt{|v^+-v^-|}\,d\#+\lambda\int_I |v-g|^2\,dt\,:\,v\in SBV(I)\Big\}.\tag{3.54}$$

Moreover, the minimum values (3.53) converge to (3.54).

*Hint*: the existence of minimum points $u_\varepsilon$ is assured by an application of the direct methods of the calculus of variations since the functional in (3.53) is coercive in $W^{2,p}(I)$ and lower semicontinuous. Use Proposition 3.22 to extract a subsequence of $u_{\varepsilon_j}$ converging in $L^1(I)$ to some $u \in SBV(I)$, noticing that the value in (3.53) is less than or equal to $\lambda\int_I |g|^2\,dt$, so that we have $\sup_{\varepsilon>0} F_\varepsilon(u_\varepsilon) < +\infty$ and $\sup_{\varepsilon>0}\|u_\varepsilon\|_{L^2(I)} < +\infty$. The minimality of $u$ and the convergence of minimum values follow from Theorem 3.3.

### 3.3.3 *Approximation of the Mumford-Shah functional by high-order perturbations*

In this section we show that it is possible to approximate the Mumford-Shah functional using the singular perturbation method introduced in the previous section.

For all $\varepsilon > 0$ let $a_\varepsilon \geq 1$ with

$$\lim_{\varepsilon\to 0+} a_\varepsilon = +\infty \qquad \text{and} \qquad \lim_{\varepsilon\to 0+} \varepsilon a_\varepsilon = 0.$$

With fixed $K > 0$ set

$$C_\varepsilon = \frac{K}{4\sqrt{\varepsilon\, a_\varepsilon}},\tag{3.55}$$

and

$$f_\varepsilon(z) = \begin{cases} z & \text{if } z \leq 1 \\ a_\varepsilon & \text{if } 1 < z < (1+C_\varepsilon)^2 \\ 0 & \text{if } z \geq (1+C_\varepsilon)^2\,. \end{cases}\tag{3.56}$$

**Theorem 3.29** *Let $I$ be a bounded open subset of* $\mathbf{R}$. *The functionals $F_\varepsilon : L^1(I) \to [0,+\infty]$ given by*

$$F_\varepsilon(u) = \begin{cases} \dfrac{1}{\varepsilon}\displaystyle\int_I f_\varepsilon(\varepsilon|u'|^2)\,dt+\varepsilon^3\int_I |u''|^2\,dt & \textit{if } u\in H^2(I) \\ +\infty & \textit{otherwise} \end{cases}\tag{3.57}$$

*$\Gamma$-converge in the $L^1(I)$-topology as $\varepsilon \to 0^+$ to the functional $F : L^1(I) \to [0, +\infty]$ given by*

$$F(u) = \begin{cases} \int_I |u'|^2\,dt + K\#(S(u)) & \text{if } u \in SBV(I) \\ +\infty & \text{otherwise.} \end{cases} \tag{3.58}$$

**Proof** We first prove the lower semicontinuity inequality. Let $u_\varepsilon \to u$ in $L^1(I)$ with $\sup_{\varepsilon>0} F_\varepsilon(u_\varepsilon, I) < +\infty$, and let

$$D_\varepsilon = \{\varepsilon|u'_\varepsilon|^2 > 1\} = \bigcup_{k\in\mathbf{N}} I^k_\varepsilon$$

in the notation of Proposition 3.24. Again, it is not restrictive to suppose that $\overline{I}^k_\varepsilon \subset I$ for all $k$. Note that for all $I^k_\varepsilon$ we have the estimate

$$\frac{1}{4\sqrt{\varepsilon}}(I^k_\varepsilon)^2 \le \int_{I^k_\varepsilon} |u_\varepsilon|\,dx \le c,$$

so that $(I^k_\varepsilon)^2 \le c\sqrt{\varepsilon}$.

We divide $D_\varepsilon$ into two families:

$$D^1_\varepsilon = \bigcup\{I^k_\varepsilon \subset D_\varepsilon : \ \varepsilon|u'_\varepsilon|^2 < (1 + C_\varepsilon)^2 \text{ on } I^k_\varepsilon\}, \qquad D^2_\varepsilon = D_\varepsilon \setminus D^1_\varepsilon.$$

If $I^k_\varepsilon \subset D^2_\varepsilon$ we have

$$\frac{1}{\varepsilon}\int_{I^k_\varepsilon} f(\varepsilon|u'_\varepsilon|^2)\,dt + \varepsilon^3\int_{I^k_\varepsilon} |u''_\varepsilon|^2\,dt \tag{3.59}$$

$$\ge 2\min_{\eta>0}\min\Big\{\frac{1}{\varepsilon}\eta a_\varepsilon + \varepsilon^3\int_0^\eta |u''|^2\,dt : u'(0) = 0,\ u'(\eta) = C_\varepsilon/\sqrt{\varepsilon}\Big\}.$$

Such a minimum can be easily computed, after remarking that it is equivalent to

$$\min_{\eta>0}\min\Big\{\eta a_\varepsilon + \int_0^\eta |u''|^2\,dt : u'(0) = 0,\ u'(\eta) = C_\varepsilon\sqrt{\varepsilon}\Big\}$$
$$= \min_{\eta>0}\min\Big\{\eta a_\varepsilon + \int_0^\eta |w'|^2\,dt : w(0) = 0,\ w(\eta) = C_\varepsilon\sqrt{\varepsilon}\Big\},$$

and see, by our choice of $C_\varepsilon$, that the minimizing pair in (3.59) is

$$\eta = \frac{\varepsilon K}{4a_\varepsilon}, \qquad u(x) = \frac{\sqrt{a_\varepsilon}}{2\varepsilon^2}x^2,$$

(up to an additive constant for $u$) which gives

$$\frac{1}{\varepsilon}\int_{I_\varepsilon^k} f(\varepsilon|u_\varepsilon'|^2)\,dt + \varepsilon^3\int_{I_\varepsilon^k}|u_\varepsilon''|^2\,dt \geq K.$$

This shows that $\#(\{k : I_\varepsilon^k \subset D_\varepsilon^2\})$ is equibounded. We can suppose that $\#(\{k : I_\varepsilon^k \subset D_\varepsilon^2\}) = N$, independent of $\varepsilon$. Since $|I_\varepsilon^k| \to 0$ as $\varepsilon \to 0^+$, we can suppose also that $D_\varepsilon^2$ shrinks to a finite set $H$. Of course, $\#(H) \leq N$.

For all $t > 0$ let $I_t = \{x \in I : \operatorname{dist}(x, H) > t\}$. For $j \in \mathbf{N}$, we define

$$f^j(s) = \begin{cases} s & \text{if } s \leq 1 \\ j & \text{if } s > 1 \end{cases}$$

and $F_\varepsilon^j$ as in (3.42) with $f^j$ in place of $f$; then, for fixed $t$ and $j$, we have, for $\varepsilon$ small enough $F_\varepsilon(u_\varepsilon, I_t) \geq F_\varepsilon^j(u_\varepsilon, I_t)$. By applying Proposition 3.25 we get that $u \in SBV(I_t)$ and

$$\liminf_{\varepsilon\to 0+} F_\varepsilon(u_\varepsilon, I_t) \geq \int_{I_t}|u'|^2 + m(j,1)\sum_{I_t\cap S(u)}\sqrt{|u^+ - u^-|}.$$

By the arbitrariness of $j$ we then have that $u \in H^1(I_t)$ for all $t > 0$, and

$$\liminf_{\varepsilon\to 0+} F_\varepsilon(u_\varepsilon, I_t) \geq \int_{I_t}|u'|^2. \tag{3.60}$$

By the arbitrariness of $t$ we obtain that $u \in SBV(I)$ and $S(u) \subset H$. On the other hand, we clearly have, for fixed $t$,

$$\liminf_{\varepsilon\to 0+} F_\varepsilon(u_\varepsilon, I\setminus\overline{I}_t) \geq \liminf_{\varepsilon\to 0+} F_\varepsilon(u_\varepsilon, D_\varepsilon^2) \geq KN \geq K\#(S(u)). \tag{3.61}$$

By (3.60), (3.61), and the arbitrariness of $t$ we get

$$\liminf_{\varepsilon\to 0+} F_\varepsilon(u_\varepsilon, I) \geq \int_I |u'|^2\,dt + K\#(S(u)).$$

Now, it suffices to construct a recovery sequence in the case $u \in SBV(-1,1)$ with $S(u) = \{0\}$, $u \in H^2((-1,1)\setminus\{0\})$ and $u'(0\pm) = 0$. The general case follows from Propositions 3.27 and 3.28.

We suppose without loss of generality that $u^+(0) > u^-(0)$. Let $z = u^+(0) - u^-(0)$, and let

$$\eta_\varepsilon = \frac{z\sqrt{\varepsilon}}{2(1+C_\varepsilon)} + \frac{(1+C_\varepsilon)\varepsilon\sqrt{\varepsilon}}{2\sqrt{a_\varepsilon}}.$$

For $\varepsilon$ small enough the following function is well-defined and belongs to the space $H^2(-1,1)$:

$$u_\varepsilon(x) = \begin{cases} u(x+\eta_\varepsilon) & \text{if } x \le -\eta_\varepsilon \\ u^-(0) + \frac{\sqrt{a_\varepsilon}}{2\varepsilon^2}(x+\eta_\varepsilon)^2 & \text{if } -\eta_\varepsilon < x < -\eta_\varepsilon + \frac{(1+C_\varepsilon)\varepsilon\sqrt{\varepsilon}}{\sqrt{a_\varepsilon}} \\ u^-(0) + \frac{(1+C_\varepsilon)^2\varepsilon}{2\sqrt{a_\varepsilon}} & \\ \quad + \frac{1+C_\varepsilon}{\sqrt{\varepsilon}}\left(x+\eta_\varepsilon - \frac{(1+C_\varepsilon)\varepsilon\sqrt{\varepsilon}}{\sqrt{a_\varepsilon}}\right) & \text{if } |x| \le \eta_\varepsilon - \frac{(1+C_\varepsilon)\varepsilon\sqrt{\varepsilon}}{\sqrt{a_\varepsilon}} \\ u^+(0) - \frac{\sqrt{a_\varepsilon}}{2\varepsilon^2}(x-\eta_\varepsilon)^2 & \text{if } \eta_\varepsilon - \frac{(1+C_\varepsilon)\varepsilon\sqrt{\varepsilon}}{\sqrt{a_\varepsilon}} < x < \eta_\varepsilon \\ u(x-\eta_\varepsilon) & \text{if } x \ge \eta_\varepsilon . \end{cases}$$

The function $u_\varepsilon$ is obtained "filling the gap of $u$" with two minimizers for the minimum problem in (3.59) joined by a steep affine function, with slope $(1 + C_\varepsilon)/\sqrt{\varepsilon}$ so that it gives no contribution to the integrals. A direct computation gives

$$\limsup_{\varepsilon\to 0+} F_\varepsilon(u_\varepsilon) = \int_I |u'|^2\,dt + K = F(u),$$

and the proof is concluded. □

### 3.3.4 *Exercises*

**Exercise 3.8** Let $g \in L^2(I)$, $\lambda > 0$, and let $(\varepsilon_j)$ be a sequence of positive numbers converging to 0. Prove that, for every sequence $(u_j)$ of minimizers of the problems

$$\min\Big\{\frac{1}{\varepsilon_j}\int_I f_{\varepsilon_j}(\varepsilon_j|u'|^2)\,dt + \varepsilon_j^3\int_I |u''|^2\,dt + \lambda\int_I |u-g|^2\,dt :\ u \in H^2(I)\Big\}$$

there exists a subsequence converging in $L^1(I)$ to a minimizer of the problem

$$\min\Big\{\int_I |u'|^2\,dt + \#(S(u)) + \lambda\int_I |u-g|^2\,dt :\ u \in SBV(I)\Big\},$$

and we have also convergence of the minimum values.

*Hint*: use the argument of Exercise 3.7

## 3.4 Non-local approximations

### 3.4.1 *Non-local approximation of the Mumford-Shah functional*

It is possible to approximate the Mumford-Shah functional by a family of "mildly non-local" functionals, of the form

$$F_\varepsilon(u) = \frac{1}{\varepsilon}\int_\Omega f\Big(\varepsilon \fint_{B_\varepsilon(x)\cap\Omega} |\nabla u(y)|^2\,dy\Big)\,dx\,, \tag{3.62}$$

defined for $u \in H^1(\Omega)$, where $f$ is a suitable non-decreasing continuous (non-convex) function, and $\fint_B$ denotes the average on $B$. These functionals are non-local in the sense that their energy density at a point $x \in \Omega$ depends on the behaviour of $u$ in the whole set $B_\varepsilon(x) \cap \Omega$.

We now prove the $\Gamma$-convergence result in the 1-dimensional case, as in the following theorem. The $n$-dimensional case will be dealt with in Chapter 5.

**Theorem 3.30** *Let $f:[0,+\infty)\to[0,+\infty)$ be a lower semicontinuous increasing function, such that $\alpha,\beta\in(0,+\infty)$ exist with*

$$\alpha=\lim_{s\to0+}\frac{f(s)}{s},\qquad \beta=\lim_{s\to+\infty}f(s), \tag{3.63}$$

*let $F_\varepsilon:L^1_{\rm loc}(\mathbf{R})\to[0,+\infty]$ be defined by*

$$F_\varepsilon(u)=\frac{1}{\varepsilon}\int_{-\infty}^{+\infty}f\Big(\frac{1}{2}\int_{x-\varepsilon}^{x+\varepsilon}|u'(y)|^2\,dy\Big)\,dx, \tag{3.64}$$

*where it is understood that*

$$f\Big(\frac{1}{2}\int_{x-\varepsilon}^{x+\varepsilon}|u'(y)|^2\,dy\Big)=\beta$$

*if $u\notin H^1(x-\varepsilon,x+\varepsilon)$. Then the family $(F_\varepsilon)$ $\Gamma$-converges as $\varepsilon\to0^+$ to the functional*

$$F(u)=\begin{cases}\alpha\displaystyle\int_{-\infty}^{+\infty}|u'|^2\,dt+2\beta\#(S(u)) & \text{if } u\in SBV_{\rm loc}(\mathbf{R})\\ +\infty & \text{otherwise}\end{cases} \tag{3.65}$$

*with respect to the $L^1_{\rm loc}(\mathbf{R})$ convergence.*

**Proof** We begin by dealing with the case

$$f(z)=\begin{cases}az & \text{if } 0\le z\le C\\ b & \text{if } z>C,\end{cases} \tag{3.66}$$

with $aC\le b$, so that $f$ is lower semicontinuous.

*Discretization of the functional $F_\varepsilon$.* If $v\in L^1_{\rm loc}(\mathbf{R})$, let

$$g(x)=f\Big(\frac{1}{2}\int_{x-\varepsilon}^{x+\varepsilon}|v'(y)|^2\,dy\Big)\qquad x\in\mathbf{R}.$$

We can write

$$F_\varepsilon(u)=\sum_{j\in\mathbf{Z}}\frac{1}{\varepsilon}\int_{2\varepsilon j}^{2\varepsilon(j+1)}g(x)\,dx=\int_0^{2\varepsilon}h(x)\,dx,$$

where

$$h(x)=\sum_{j\in\mathbf{Z}}\frac{1}{\varepsilon}g(x+2\varepsilon j).$$

By the Mean Value Theorem we find $\xi\in(0,2\varepsilon)$ such that

$$\int_0^{2\varepsilon} h(x)\,dx \geq 2\varepsilon\, h(\xi);$$

hence, we have

$$F_\varepsilon(v) \geq \sum_{j\in\mathbf{Z}} 2\,f\Big(\frac{1}{2}\int_{\xi+\varepsilon(2j-1)}^{\xi+\varepsilon(2j+1)} |v'|^2\,dy\Big). \tag{3.67}$$

*Proof of the lower semicontinuity inequality.* With fixed $u \in L^\infty(\mathbf{R})$, let $u_k \to u$ in $L^1_{\rm loc}(\mathbf{R})$ with $\liminf_k F_{\varepsilon_k}(u_k) < +\infty$. We can suppose that $\sup_k F_{\varepsilon_k}(u_k) < +\infty$. Moreover, by a truncation argument we can suppose also $\|u_k\|_\infty \leq \|u\|_\infty$, and, up to a translation, that taking $\varepsilon = \varepsilon_k$ and $v = u_k$ in (3.67) we have $\xi = 0$

For all $k$ we divide $\mathbf{Z}$ into two sets of indices:

$$G_k = \{j \in \mathbf{Z} : \int_{\varepsilon_k(2j-1)}^{\varepsilon_k(2j+1)} |u_k'|^2\,dy \leq C\},$$

$$B_k = \{j \in \mathbf{Z} : \int_{\varepsilon_k(2j-1)}^{\varepsilon_k(2j+1)} |u_k'|^2\,dy > C\}.$$

We then have by (3.67)

$$\begin{aligned} F_{\varepsilon_k}(u_k) &\geq \sum_{j\in B_k} 2f\Big(\frac{1}{2}\int_{\varepsilon_k(2j-1)}^{\varepsilon_k(2j+1)} |u_k'|^2\,dy\Big) + \sum_{j\in G_k} 2f\Big(\frac{1}{2}\int_{\varepsilon_k(2j-1)}^{\varepsilon_k(2j+1)} |u_k'|^2\,dy\Big) \\ &= 2b\,\#(B_k) + a\sum_{j\in G_k}\int_{\varepsilon_k(2j-1)}^{\varepsilon_k(2j+1)} |u_k'|^2\,dy. \end{aligned}$$

Note that $\#(B_k) \leq c$ independent of $k$. We define

$$v_k(x) = \begin{cases} u_k(x) & \text{if } x \in 2\varepsilon_k G_k + (-\varepsilon_k, \varepsilon_k), \\ u_k^-((2j-1)\varepsilon_k) & \text{if } x \in 2\varepsilon_k B_k + [-\varepsilon_k, \varepsilon_k]. \end{cases}$$

Note that $v_k$ is discontinuous at most at the points of the form $(2j+1)\varepsilon_k$ with $j \in B_k$; moreover,

$$\|u_k - v_k\|_{L^1(\mathbf{R})} \leq \#(B_k)4\varepsilon_k\|u\|_\infty \to 0,$$

so that in particular $v_k \to u$ in $L^1_{\rm loc}(\mathbf{R})$

$$\#(S(v_k)) \leq \#(B_k), \qquad \int_{-\infty}^{+\infty} |v_k'|^2\,dx = \sum_{j\in G_k}\int_{\varepsilon_k(2j-1)}^{\varepsilon_k(2j+1)} |u_k'|^2\,dy.$$

Hence,

$$a\int_{-\infty}^{+\infty}|v_k'|^2\,dx+2b\,\#(S(v_k))\le F_{\varepsilon_k}(u_k)\le c<+\infty.$$

We can apply Theorem 2.3 on each bounded interval of $\mathbf{R}$ and obtain that $u\in SBV_{\rm loc}(\mathbf{R})$. Moreover, by the lower semicontinuity of the Mumford-Shah functional, for all bounded intervals $I$

$$\begin{aligned}a\int_I|u'|^2\,dx+2b\#(S(u)\cap I)&\le\liminf_k\Bigl(a\int_I|v_k'|^2\,dx+2b\#(S(v_k)\cap I)\Bigr)\\&\le\liminf_k F_{\varepsilon_k}(u_k).\end{aligned}$$

This proves the $\Gamma$-liminf inequality if $u\in L^\infty(\mathbf{R})$.

In the general case if $u_j\to u$ in $L^1_{\rm loc}(\mathbf{R})$ then for all $T>0$ we have $u_{T,j}:=(-T)\vee(T\wedge u_j)\to u_T:=(-T)\vee(T\wedge u)$ in $L^1_{\rm loc}(\mathbf{R})$. Hence,

$$a\int_{-\infty}^{+\infty}|u_T'|\,dt+2b\#(S(u_T))\le\liminf_j F_{\varepsilon_j}(u_{T,j})\le\liminf_j F_{\varepsilon_j}(u_j).$$

The required inequality follows taking the supremum for $T>0$.

*Construction of the recovery sequence.* It suffices to consider the case when $u\in SBV_{\rm loc}(\mathbf{R})$ satisfies $u'\in L^2(\mathbf{R})$ and $\#(S(u))<+\infty$. We can take simply $u_\varepsilon=u$ for all $\varepsilon>0$. To simplify calculations, we restrict to the case when $S(u)=\{0\}$. Note, after changing the order of integration, that

$$\frac{1}{2\varepsilon}\int_{-\infty}^{+\infty}\int_{x-\varepsilon}^{x+\varepsilon}|u'(t)|^2\,dt\,dx=\int_{-\infty}^{+\infty}|u'(t)|^2\,dt.$$

Hence, since $(1/\varepsilon)f(\varepsilon\xi)\to a\xi$ for all $\xi\ge0$, and we have $f(t)\le ct$ for some $c>0$ and for all $t\ge0$, we can apply the Lebesgue Dominated Convergence Theorem to the family

$$x\mapsto\frac{1}{\varepsilon}f\Bigl(\frac12\int_{x-\varepsilon}^{x+\varepsilon}|u'(t)|^2\,dt\Bigr)\,\chi_{[-\varepsilon,\varepsilon]}(x),$$

as $\varepsilon\to0^+$, obtaining

$$\lim_{\varepsilon\to0+}\frac{1}{\varepsilon}\int_{(-\infty,-\varepsilon)\cup(\varepsilon,+\infty)}f\Bigl(\frac12\int_{x-\varepsilon}^{x+\varepsilon}|u'(t)|^2\,dt\Bigr)\,dx=a\int_{-\infty}^{+\infty}|u'|^2\,dx.$$

On the other hand, we trivially have

$$\frac{1}{\varepsilon}\int_{[-\varepsilon,\varepsilon]}f\Bigl(\frac12\int_{x-\varepsilon}^{x+\varepsilon}|u'(t)|^2\,dt\Bigr)\,dx\le b,$$

so that

$$\limsup_{\varepsilon\to0+}F_\varepsilon(u)\le a\int_{-\infty}^{+\infty}|u'|^2\,dx+b=F(u).$$

Clearly, the same proof holds if $\#(S(u))<+\infty$.

*Proof in the case of a general f.* Define

$$F_\varepsilon(u,I) = \frac{1}{\varepsilon}\int_I f\Big(\frac{1}{2}\int_{x-\varepsilon}^{x+\varepsilon} |u'(y)|^2\,dy\Big)\,dx\,, \tag{3.68}$$

for $I \subset \mathbf{R}$ and $u \in L^1_{\mathrm{loc}}(\mathbf{R})$. It can be easily seen that for $u \in SBV_{\mathrm{loc}}(\mathbf{R})$ with $\#(S(u)) < +\infty$ and $u' \in L^2(\mathbf{R})$, the set function

$$\mu(I) = \inf\{\liminf_j F_{\varepsilon_j}(u_j, I) : \ u_j \to u \text{ in } L^1_{\mathrm{loc}}(\mathbf{R}),\ \varepsilon_j \to 0^+\}$$

defines a superadditive set function on open sets with disjoint and compact closures in $\mathbf{R}$. If $(a_i)$, $(b_i)$ are sequences of positive numbers such that $\sup_i a_i = \alpha$, $\sup_i b_i = \beta$, and $\max\{a_i t, b_i\} \le f(t)$ for all $t$, then, proceeding as above, we get

$$\mu(I) \ge a_i \int_I |u'|^2\,dt + 2b_i \#(S(u)\cap I).$$

By Proposition 1.16 we get that

$$\mu(I) \ge \alpha \int_I |u'|^2\,dt + 2\beta\#(S(u)\cap I),$$

i.e., taking $I = \mathbf{R}$, the inequality

$$\alpha \int_{-\infty}^{+\infty} |u'|^2\,dt + 2\beta\#(S(u)) \le \Gamma\text{-}\liminf_{\varepsilon\to 0+} F_\varepsilon(u)$$

holds as required.

Conversely, for all $\eta > 0$ there exists $C_\eta > 0$ such that $(\alpha+\eta)C_\eta \le \beta$ and the function

$$\tilde f(z) = \begin{cases} (\alpha+\eta)z & \text{if } 0 \le z \le C_\eta \\ \beta & \text{if } z > C_\eta, \end{cases}$$

satisfies $\tilde f \ge f$. Then we get by comparison

$$\Gamma\text{-}\limsup_{\varepsilon\to 0+} F_\varepsilon(u) \le (\alpha+\eta)\int_{-\infty}^{+\infty} |u'|^2\,dt + 2\beta\#(S(u)\cap I),$$

and the thesis follows by the arbitrariness of $\eta > 0$. □

### 3.4.2 *Exercises*

**Exercise 3.9** Let $\varepsilon > 0$ and $I \subset \mathbf{R}$ be open and bounded. Prove that the functional

$$F(u) = \int_I f\Big(\int_{x-\varepsilon}^{x+\varepsilon} |u'|^2\,dt\Big)\,dx$$

defined as in Theorem 3.30 is lower semicontinuous on $L^1_{\mathrm{loc}}(\mathbf{R})$ if and only if $f$ is lower semicontinuous and increasing.

*Hint*: use Fatou's Lemma and the weak lower semicontinuity of the $L^2$ norm of the gradient to prove the lower semicontinuity of $F_\varepsilon$. To prove the necessity of the lower semicontinuity of $f$ use strong approximations on affine functions; to prove the necessity that $f$ be increasing approximate, e.g., the function $u(x) = ax_1$ by $u_j(x) = ax_1 + t\sin(jx_1)$ ($t \geq 0$).

**Exercise 3.10** Let $f$ be as in Theorem 3.30. Prove that the functionals

$$\widetilde{F}_\varepsilon(u) = \begin{cases} \dfrac{1}{\varepsilon}\displaystyle\int_{-\infty}^{+\infty} f\Big(\frac{1}{2}\int_{x-\varepsilon}^{x+\varepsilon} |u'(y)|^2\,dy\Big)\,dx & u \in H^1_{\rm loc}(\mathbf{R}) \\ +\infty & \text{otherwise} \end{cases} \tag{3.69}$$

$\Gamma$-converge as $\varepsilon \to 0^+$ to the functional $F$ defined in (3.65) with respect to the $L^1_{\rm loc}(\mathbf{R})$ convergence.

*Hint*: by Theorem 3.30 we obtain by comparison the lower semicontinuity inequality. To build a recovery sequence for $u \in SBV_{\rm loc}(\mathbf{R})$ with $\#(S(u)) < +\infty$ and $S(u) = \{0\}$ take any $u_\varepsilon \in H^1_{\rm loc}(\mathbf{R})$ with $u_\varepsilon(x) = u(x)$ if $|x| > \varepsilon^2$.

**Exercise 3.11** With fixed $\varepsilon > 0$, let $\widetilde{F}_\varepsilon$ be defined in (3.69). Prove that the lower semicontinuous envelope of $\widetilde{F}_\varepsilon$ is the functional $F_\varepsilon$ defined in Theorem 3.30.

### 3.4.3 *Non-local approximation of free-discontinuity problems*

In this section we show that more general surface energies can be obtained by non-local approximation of the type considered in the previous section, by taking into account functionals of the form

$$F_\varepsilon(u) = \frac{1}{\varepsilon}\int_\Omega f_\varepsilon\Big(\varepsilon \fint_{B_\varepsilon(x)} |\nabla u|^2\,dy\Big)\,dx, \qquad u \in H^1(\Omega), \tag{3.70}$$

with varying integrands $f_\varepsilon$. Note that if $f_\varepsilon = f$ then we may obtain in the limit only a Mumford-Shah functional.

In what follows, fixed a non-decreasing function $g : [0,+\infty) \to [0,+\infty)$ with

$$g(0) = 0, \qquad \inf\{g(x) : x > 0\} =: c_g > 0\,, \tag{3.71}$$

we will consider a new function $\varphi$ defined by

$$\varphi(z) = \inf\Big\{\int_{-\infty}^{+\infty} g\Big(\int_{x-1}^{x+1} |u'(t)|^2\,dt\Big)\,dx : u(-\infty) = 0,\ u(+\infty) = z\Big\} \tag{3.72}$$

where the infimum is taken over all functions in $H^1_{\rm loc}(\mathbf{R})$. The meaning of the conditions at $\pm\infty$ is understood as the existence of the corresponding limits. It can be easily checked that the function $\varphi$ is lower semicontinuous and subadditive, and $\inf_{z>0}\varphi(z) > 0$. The function $\varphi$ can be interpreted as an asymptotic non-local least transition energy between two phases of distance $z$. Note that in the case $f_\varepsilon = f$, as in the previous section, the limit function $g$ is a constant $\lambda$

(except at 0 where $g(0) = 0$), so that $\varphi(z) = 2\lambda$, and the minimizers $u$ undergo a brutal transition at one point. This phenomenon does not take place for a general $g$, and taking care of the possible behaviours of minimizers for the problem defining $\varphi$ is one of the reasons why this case is trickier that the previous one.

**Remark 3.31** Note that in the definition of $\varphi$ it suffices to consider functions $u$ with derivatives of compact support. In fact, if we define

$$w_u(t) := \int_{t-1}^{t+1} |u'(x)|^2\,dx, \qquad G(u) := \int_{-\infty}^{+\infty} g(w_u(t))\,dt, \tag{3.73}$$

and $G(u) < +\infty$, then $|\mathrm{spt}\, w_u| \le c_g^{-1} G(u)$, and also $|\mathrm{spt}\, u'| \le c_g^{-1} G(u) - 2$. Note that $\mathrm{spt}\, u'$ cannot contain more than $|\mathrm{spt}\, w_u|/2$ points of mutual distance greater than 2, so that it is not restrictive to assume, up to eliminating some bounded intervals where $u$ is constant, that

$$\mathrm{spt}\, w_u \subseteq [0, c_g^{-1}\varphi(z) + 1]\,. \tag{3.74}$$

Hence, we can suppose that $u(x) = 0$ for $x < 1$, and $u(x) = z$ for $x > c_g^{-1}\varphi(z)$.

Moreover, by considering $u_z(t) = \mathrm{sign}\, z\,((0 \vee |z|t) \wedge |z|)$ as a test function, we deduce that

$$\varphi(z) \le \int_{-\infty}^{+\infty} g\Big(\int_{x-1}^{x+1} |u_z'(t)|^2\,dt\Big)\,dx \le 3\,g(z^2)\,,$$

so that we can assume that

$$\mathrm{spt}\, w_u \subseteq [0, 3c_g^{-1} g(z^2) + 1], \tag{3.75}$$

in place of (3.74).

**Theorem 3.32** *Let $g : [0,+\infty) \to [0,+\infty)$ be a lower semicontinuous increasing function satisfying* (3.71). *Define*

$$f_\varepsilon(\xi) = \begin{cases} \xi & \text{if } 0 \le \xi \le c_g \\ g(\varepsilon\xi) & \text{if } \xi > c_g, \end{cases}$$

*and*

$$F_\varepsilon(u) = \frac{1}{\varepsilon}\int_{-\infty}^{+\infty} f_\varepsilon\Big(\int_{x-\varepsilon}^{x+\varepsilon} |u'(t)|^2 dt\Big)\,dx, \qquad u \in H^1_{\mathrm{loc}}(\mathbf{R})$$

*(extended to $+\infty$ on $L^1_{\mathrm{loc}}(\mathbf{R}) \setminus H^1_{\mathrm{loc}}(\mathbf{R})$). Then the $\Gamma$-limit in $L^1_{\mathrm{loc}}(\mathbf{R})$ of $F_\varepsilon$ as $\varepsilon \to 0^+$ exists, and it equals the functional $F_\varphi$ defined by*

$$F_\varphi(u) := 2\int_{-\infty}^{+\infty} |u'|^2\,dx + \sum_{x\in S(u)} \varphi(u^+(x) - u^-(x)), \qquad u \in SBV_{\mathrm{loc}}(\mathbf{R}),$$

*(extended to $+\infty$ on $L^1_{\mathrm{loc}}(\mathbf{R}) \setminus SBV_{\mathrm{loc}}(\mathbf{R})$), where $\varphi$ is the "transition energy density" defined by* (3.72).

**Proof** Let $(\varepsilon_j)$ tend to 0, and let $u_j \to u$ in $L^1_{\rm loc}(\mathbf{R})$. With a slight abuse of notation we set $F_j := F_{\varepsilon_j}$. We have to check that

$$F_\varphi(u) \leq \liminf_j F_j(u_j). \tag{3.76}$$

We can suppose that $\sup_j F_j(u_j) < +\infty$. For each $j \in \mathbf{N}$, define $\psi_j(x) = f_{\varepsilon_j}\Big(\int_{x-\varepsilon_j}^{x+\varepsilon_j} |u_j'(t)|^2\,dt\Big)$, so that

$$F_j(u_j) = \int_{-\infty}^{+\infty} \frac{1}{\varepsilon_j}\psi_j(x)\,dx = \sum_{k\in\mathbf{Z}} \int_{2k\varepsilon_j-\varepsilon_j}^{2k\varepsilon_j+\varepsilon_j} \frac{1}{\varepsilon_j}\psi_j(x)\,dx = \frac{1}{\varepsilon_j}\int_{-\varepsilon_j}^{\varepsilon_j} \Psi_j(x)\,dx,$$

where $\Psi_j(x) := \sum_{k\in\mathbf{Z}} \psi_j(x + 2k\varepsilon_j)$. By the Mean Value Theorem, we can find $t_j \in (-\varepsilon_j, \varepsilon_j)$ such that

$$F_j(u_j) \geq 2\Psi_j(t_j) = 2\sum_{k\in\mathbf{Z}} f_{\varepsilon_j}\Big(\int_{(2k-1)\varepsilon_j}^{(2k+1)\varepsilon_j} |u_j'(t-t_j)|^2\,dt\Big).$$

Note that, since $t_j \to 0$ the functions $v_j(t) = u_j(t - t_j)$ still converge to $u$, and $F_j(u_j) = F_j(v_j)$. Hence, we can suppose $t_j = 0$ for all $j$, and

$$F_j(u_j) \geq 2\sum_{k\in\mathbf{Z}} f_{\varepsilon_j}\Big(\int_{(2k-1)\varepsilon_j}^{(2k+1)\varepsilon_j} |u_j'(t)|^2\,dt\Big).$$

Define

$$\xi_j(k) := \int_{(2k-1)\varepsilon_j}^{(2k+1)\varepsilon_j} |u_j'(t)|^2\,dt, \qquad k \in \mathbf{Z},$$

$$G_j := \{k \in \mathbf{Z} : \ 2\xi_j(k) \leq c_g\}, \qquad B_j := \{k \in \mathbf{Z} : \ 2\xi_j(k) > c_g\}.$$

Note that for all $x \in \mathbf{R} \setminus (2\varepsilon_j B_j + [-2\varepsilon_j, 2\varepsilon_j])$ we have

$$\int_{x-\varepsilon_j}^{x+\varepsilon_j} |u_j'(t)|^2\,dt \leq c_g.$$

Moreover, $f_{\varepsilon_j}(\xi_j(k)) \geq c_g/2$ for all $k \in B_j$, so that

$$\#(B_j) \leq \frac{2}{c_g}\sum_{k\in B_j} f_{\varepsilon_j}(\xi_j(k)) \leq \frac{1}{c_g}\sup_j F_j(u_j).$$

Hence, we can suppose that $\#(B_j) = N$, independent of $j$. If we write $B_j = \{k_j^1, \ldots, k_j^N\}$ with $k_j^1 < k_j^2 < \ldots < k_j^N$, then we can suppose that there exist indices $i_1, i_2$ such that $\varepsilon_j k_j^i \to -\infty$ for $i < i_1$, $\varepsilon_j k_j^i \to +\infty$ for $i > i_2$, and $2\varepsilon_j k_j^i \to x^i \in \mathbf{R}$ for $i_1 \leq i \leq i_2$. Let $S = \{x^i : i_1 \leq i \leq i_2\}$ (which is not empty if $i_1 \leq i_2$). Since

$$F_j(u_j) \geq \frac{1}{\varepsilon_j} \int_{\mathbf{R}\setminus(2\varepsilon_j B_j + [-2\varepsilon_j, 2\varepsilon_j])} \int_{x-\varepsilon_j}^{x+\varepsilon_j} |u_j'(t)|^2 \, dt \, dx$$
$$\geq 2 \int_{\mathbf{R}\setminus(2\varepsilon_j B_j + [-2\varepsilon_j, 2\varepsilon_j])} |u_j'(t)|^2 \, dt$$

(the last inequality is obtained by changing the integration order), we have that fixed $\eta > 0$ there exists $j(\eta)$ such that the sequence $(u_j)_{j \geq j(\eta)}$ is equibounded in $H^1([-\frac{1}{\eta}, \frac{1}{\eta}] \setminus (S + [-\eta, \eta]))$, and

$$\int_{[-\frac{1}{\eta}, \frac{1}{\eta}]\setminus(S+[-\eta,\eta])} |u_j'|^2 \, dt \leq \sum_{k \in G_j} \int_{(2k-1)\varepsilon_j}^{(2k+1)\varepsilon_j} |u_j'|^2 \, dt$$

for $j \geq j(\eta)$. It follows that $u \in SBV_{\rm loc}(\mathbf{R})$, $S(u) \subset S$, and

$$\int_{-\infty}^{+\infty} |u'|^2 \, dt \leq \liminf_j \sum_{k \in G_j} \int_{(2k-1)\varepsilon_j}^{(2k+1)\varepsilon_j} |u_j'|^2 \, dt \, .$$

By the local nature of the arguments in the proofs below it will not be restrictive to suppose $x^1 = \ldots = x^N = 0$. Moreover, we can suppose that $u_j$ is constant in $[(2k_j^1 - 3)\varepsilon_j, (2k_j^1 - 1)\varepsilon_j]$ and in $[(2k_j^N + 1)\varepsilon_j, (2k_j^N + 3)\varepsilon_j]$. This is not restrictive, up to substituting $u_j$ by a function $v_j$ constant on these intervals, with $v_j' = u_j'$ elsewhere, and coinciding with $u_j$ for $t < (2k_j^1 - 3)\varepsilon_j$. Clearly $F_j(v_j) \leq F_j(u_j)$, and still $v_j \to u$ since

$$\|u_j - v_j\|_\infty \leq \int_{-\infty}^{+\infty} |u_j' - v_j'| \, dt \leq 2\sqrt{c_g \varepsilon_j}$$

(using Hölder's inequality). We can split $F_j(u_j)$ into three integrals, and we then have

$$F_j(u_j) \geq 2 \int_{-\infty}^{(2k_j^1-2)\varepsilon_j} |u_j'|^2 \, dt + 2 \int_{(2k_j^N+2)\varepsilon_j}^{+\infty} |u_j'|^2 \, dt$$
$$+ \frac{1}{\varepsilon_j} \int_{(2k_j^1-2)\varepsilon_j}^{(2k_j^N+2)\varepsilon_j} f_{\varepsilon_j}\Big( \int_{x-\varepsilon_j}^{x+\varepsilon_j} |u_j'|^2 \, dt \Big) \, dx.$$

To treat the last term, we can suppose that $u_j$ is monotone on the interval $[(2k_j^1 - 2)\varepsilon_j, (2k_j^N + 2)\varepsilon_j]$ and constant on the intervals $[(2k-1)\varepsilon_j, (2k+1)\varepsilon_j]$ with $k \in G_j$, $k_j^1 < k < k_j^N$. In fact, if this does not occur, then, using the same argument as above, we can substitute $u_j$ by a function $v_j$ enjoying these properties, with $v_j \to u$ and $F_j(v_j) \le F_j(u_j)$. Hence, we can also suppose that $k_j^{i+1} - k_j^i \le 2$ for $i = 1, \dots, N-1$ so that $k_j^N - k_j^1 \le 2N+2$. Finally, by a translation argument it is not restrictive to suppose that $u_j(0) = 0 = u^-(0)$ for all $j$ (we tacitly use the continuous representatives of Sobolev functions throughout the section).

Define for all $K > 0$, $z \in \mathbf{R}$

$$\varphi_\varepsilon^K(z) = \inf\Big\{\int_0^K f_\varepsilon\Big(\frac{1}{\varepsilon}\int_{x-1}^{x+1} |v'|^2\,dt\Big)\,dx : v(0) = 0,\ v(K) = z\Big\}. \tag{3.77}$$

By the lower semicontinuity of $f_\varepsilon$ the function $\varphi_\varepsilon^K$ is itself lower semicontinuous. Note that $f_{\varepsilon_j}(\cdot/\varepsilon_j)$ converges increasingly to $g$; hence, $\varphi_{\varepsilon_j}^K$ converges increasingly to

$$\varphi^K(z) = \inf\Big\{\int_0^K g\Big(\int_{x-1}^{x+1} |v'|^2\,dt\Big)\,dx : v(0) = 0,\ v(K) = z\Big\}.$$

It can be easily checked that $\varphi^K(z) = \varphi(z)$ for $K$ large, and that $\varphi(z) \ge 2c_g$ if $z \ne 0$.

Let $K \ge 4N+4$ be fixed. Define

$$v_j(x) = \begin{cases} u_j(x + (2k_j^1 - 2)\varepsilon_j) & \text{if } x \le 0 \\ u_j(x + (2k_j^N + 2)\varepsilon_j) & \text{if } x > 0. \end{cases}$$

Note that

$$\frac{1}{\varepsilon_j}\int_{(2k_j^1-2)\varepsilon_j}^{(2k_j^N+2)\varepsilon_j} f_{\varepsilon_j}\Big(\int_{x-\varepsilon_j}^{x+\varepsilon_j} |u_j'|^2\,dt\Big)\,dx \ge \varphi_{\varepsilon_j}^K(v_j^+(0) - v_j^-(0)),$$

using $v(t) = u_j(t/\varepsilon_j)$ as a test function in (3.77), and performing a change of variables. Moreover, since we suppose $N \ge 1$ (otherwise there is nothing to prove), we have

$$\frac{1}{\varepsilon_j}\int_{(2k_j^1-2)\varepsilon_j}^{(2k_j^N+2)\varepsilon_j} f_{\varepsilon_j}\Big(\int_{x-\varepsilon_j}^{x+\varepsilon_j} |u_j'|^2\,dt\Big)\,dx \ge 2\,g\Big(\int_{2k_j^1\varepsilon_j-\varepsilon_j}^{2k_j^1\varepsilon_j+\varepsilon_j} |u_j'|^2\,dt\Big) \ge 2c_g,$$

so that

$$F_j(u_j) \ge 2\int_{-\infty}^{+\infty} |v_j'|^2\,dx + \varphi_{\varepsilon_j}^K(v_j^+(0) - v_j^-(0)) \vee 2c_g.$$

Note that $v_j \to u$ in $L^1_{\rm loc}(\mathbf{R})$, and that $(v_j)$ is bounded in $H^1((-T,0)\cup(0,T))$ for all $T > 0$, so that actually $v_j \to u$ in $L^\infty_{\rm loc}(\mathbf{R})$. In particular $v_j^\pm(0) \to u^\pm(0)$, and $v_j^+(0) - v_j^-(0) \to u^+(0) - u^-(0)$.

Since $\varphi_{\varepsilon_j}^K$ is a sequence of lower semicontinuous functions converging increasingly to $\varphi$ we have $\varphi(z) \le \liminf_j \varphi_{\varepsilon_j}^K(z_j)$ for all $z_j \to z$. Hence,

$$\liminf_j F_j(u_j) \geq \liminf_j 2\int_{-\infty}^{+\infty} |v_j'|^2\,dt + \liminf_j \varphi^K_{\varepsilon_j}(v_j^+(0) - v_j^-(0))$$
$$\geq 2\int_{-\infty}^{+\infty} |u'|^2\,dt + \varphi(u^+(0) - u^-(0)) = F_\varphi(u),$$

that is (3.76).

It now remains to find a recovery sequence for $F_\varphi(u)$ when $u \in SBV_{\rm loc}(\mathbf{R})$, $u' \in L^2(\mathbf{R})$ and $\#(S(u)) < +\infty$. It is not restrictive to suppose $S(u) = \{0\}$ since all the arguments are local. Moreover, we can suppose $u^-(0) = 0$, $u^+(0) = z$. Fix $K$ such that $\varphi(z) = \varphi^K(z)$. For all $\varepsilon > 0$ there exists $\tilde{u}_\varepsilon \in H^1_{\rm loc}(\mathbf{R})$ such that $\tilde{u}_\varepsilon(x) = 0$, for $x \leq 0$, $\tilde{u}_\varepsilon(x) = z$, for $x \geq K\varepsilon$, and

$$\frac{1}{\varepsilon}\int_{-\infty}^{+\infty} f_\varepsilon\Big(\int_{x-\varepsilon}^{x+\varepsilon} |\tilde{u}_\varepsilon'|^2\,dt\Big)\,dx = \varphi^K_\varepsilon(z).$$

Define

$$u_\varepsilon(x) = \begin{cases} u(x+2\varepsilon) & \text{if } x \leq -2\varepsilon \\ \tilde{u}_\varepsilon(x) & \text{if } -2\varepsilon < x < K\varepsilon + 2\varepsilon \\ u(x - 2\varepsilon - K\varepsilon) & \text{if } x \geq K\varepsilon + 2\varepsilon. \end{cases}$$

We have $u_\varepsilon \to u$, and $\lim\limits_{\varepsilon\to 0+} F_\varepsilon(u_\varepsilon) = 2\int_{-\infty}^{+\infty} |u'|^2\,dt + \lim\limits_{\varepsilon\to 0+} \varphi^K_\varepsilon(z) = F_\varphi(u)$, as required. □

### 3.4.4 *Exercises*

**Exercise 3.12** Prove that in the statement of Theorem 3.32 we can define $f_\varepsilon$ also as

$$f_\varepsilon(\xi) = \begin{cases} \xi & \text{if } 0 \leq \xi \leq C \\ g(\varepsilon\xi) & \text{if } \xi > C, \end{cases}$$

for any $C \leq c_g$.

**Exercise 3.13** Recover the result of the previous section as a corollary of Theorem 3.32.

*Hint.* Let $f$ be a function as in the previous section, and let $\tilde{f}(t) = 2f(t/2)$. Rewrite the functionals $F_\varepsilon$ of the previous section in terms of $\tilde{f}$. For fixed $\eta > 0$ let $0 < C \leq 1$ satisfy

$$(1-\eta)t \leq \tilde{f}(t) \leq (1+\eta)t \text{ if } 0 \leq t \leq C,$$

$$2(\lambda - \eta) \leq \tilde{f}(t) \text{ if } t \geq 1/C.$$

Define, for $\sigma > 0$

$$g^\sigma(t) = \begin{cases} 0 & \text{if } t = 0 \\ (1-\eta)C & \text{if } 0 < t \leq \sigma \\ 2(\lambda - \eta) & \text{if } t > \sigma, \end{cases} \qquad g^+(t) = \begin{cases} 0 & \text{if } t = 0 \\ 2\lambda & \text{if } t > 0, \end{cases}$$

$$f_\varepsilon^\sigma(t) = \begin{cases} (1-\eta)t & \text{if } 0 \le t \le C \\ g^\sigma(\varepsilon t) & \text{if } t > C, \end{cases} \qquad f_\varepsilon^+(t) = \begin{cases} (1+\eta)t & \text{if } 0 \le t \le C \\ 2(1+\eta)\lambda & \text{if } t > C, \end{cases}$$

Apply Theorem 3.32 with $f_\varepsilon = \frac{1}{(1+\eta)} f_\varepsilon^+$ and $f_\varepsilon = \frac{1}{(1-\eta)} f_\varepsilon^\sigma$, noting that $f_\varepsilon^\sigma \le \tilde{\le} f^+\varepsilon$. Conclude the proof by a comparison argument and the arbitrariness of $\sigma$.

**Exercise 3.14** Prove that if

$$g(t) = \begin{cases} at + b & \text{if } t > 0 \\ 0 & \text{if } t = 0, \end{cases}$$

then

$$\varphi(z) = \begin{cases} 2\sqrt{2ab}|z| + 2b & \text{if } z \ne 0 \\ 0 & \text{if } z = 0. \end{cases}$$

**Exercise 3.15** Let $\vartheta : \mathbf{R} \to \mathbf{R}$ be a subadditive lower semicontinuous function, increasing on $[0, +\infty)$, with $\vartheta(0) = 0$. Let $g(t) = \frac{1}{2}\vartheta(\sqrt{2t})$ for $t \ge 0$. Then prove that $\varphi(z) \ge \vartheta(|z|)$.

*Hint:* use the same "discretization argument" as in the first part of the proof of Theorem 3.32, and the subadditivity of $\vartheta$.

**Exercise 3.16** Let $\vartheta, g$ be as above. Prove that $\varphi(z) \le 2\,\vartheta(|z|)$.

*Hint:* take $u(t) = \operatorname{sign} z\,\big((0 \vee |z|t/2) \wedge |z|\big)$ as a test function in the definition of $\varphi$.

## 3.5 Finite-difference approximation of free-discontinuity problems

In this section we study an approximation procedure via functionals depending on measurable functions through some difference quotients. In particular we can apply these results to functionals of the form

$$F_\varepsilon(u) = \frac{1}{\varepsilon}\int_I f\Big(\frac{(u(t+\varepsilon) - u(t))^2}{\varepsilon}\Big)\,dt, \tag{3.78}$$

defined for $u \in L^1_{\rm loc}(\mathbf{R})$, where $I \subset \mathbf{R}$ is an interval and $f$ is an increasing function differentiable at 0, and we can show that again this family Γ-converges to the Mumford-Shah functional, in a sense specified below. This result will be used to obtain an $n$-dimensional approximation in Chapter 5.

For the sake of convenience and generality we prove the Γ-convergence in several steps.

**Theorem 3.33** *Let $h : [0, +\infty) \to [0, +\infty)$ be a convex function with $h(0) = 0$, and let $\varphi : [0, +\infty) \to [0, +\infty)$ be a lower semicontinuous subadditive function satisfying*

$$\lim_{t\to+\infty} \frac{h(t)}{t} = \lim_{t\to 0+} \frac{\varphi(t)}{t} = +\infty\,. \tag{3.79}$$

*Let $C > 0$, for each $\varepsilon > 0$ let $f_\varepsilon : [0. +\infty) \to [0, +\infty)$ be the function defined by*

$$f_\varepsilon(z) = \begin{cases} \varepsilon\, h(\sqrt{z/\varepsilon}) & \text{if } z \le C \\ \varphi(\sqrt{\varepsilon z}) & \text{if } z > C\,, \end{cases} \tag{3.80}$$

*for each* $u \in L^1_{\rm loc}(\mathbf{R})$ *and* $I \subset \mathbf{R}$ *open set, let*

$$F_\varepsilon(u, I) = \frac{1}{\varepsilon}\int_I f_\varepsilon\Big(\frac{(u(t+\varepsilon) - u(t))^2}{\varepsilon}\Big)\,dt,. \tag{3.81}$$

*Then if* $I$ *is an open interval the family* $F_\varepsilon(\cdot, I)$ $\Gamma$*-converges as* $\varepsilon \to 0^+$ *to the functional* $F(\cdot, I)$*, with respect to the* $L^1_{\rm loc}(\mathbf{R})$*-convergence, where*

$$F(u, I) = \begin{cases} \displaystyle\int_I h(|u'|)\,dt + \sum_{S(u)\cap I} \varphi(|u^+ - u^-|) & \text{if } u \in SBV_{\rm loc}(I) \\ +\infty & \text{otherwise} \end{cases} \tag{3.82}$$

*is defined for* $u \in L^1_{\rm loc}(\mathbf{R})$.

**Remark 3.34** Theorem 3.33 does not hold for arbitrary open sets $I$. For instance, take $I = \mathbf{R} \setminus \{0\}$. Since we have in this case $F_\varepsilon(u, I) = F_\varepsilon(u, \mathbf{R})$ we should have also $F(u, I) = F(u, \mathbf{R})$, which is not the case if $0 \in S(u)$.

**Remark 3.35** If we take $h(z) = az^2$, $\varphi = b$ constant, and $C = b/a$ then we have $f_\varepsilon(z) = f(z) = \min\{az, b\}$ and we obtain as limit the Mumford-Shah functional $a\int_I |u'|^2\,dt + b\#(S(u))$.

To prove Theorem 3.33 we need some preliminary results. A crucial step will be the construction of suitable piecewise constant or piecewise affine functions. To this purpose we need the following lemma. If $v \in L^1_{\rm loc}(\mathbf{R})$ we use the notation

$$T^\varepsilon_y v(x) = v\Big(y + \Big[\frac{x-y}{\varepsilon}\Big]\Big),$$

that gives a piecewise constant function, with constant value $v(y + \varepsilon k)$ on the interval $y + \varepsilon(k, k+1)$, $k \in \mathbf{Z}$. Note that $T^\varepsilon_{y_\varepsilon} v = T^\varepsilon_y$.

**Lemma 3.36** *Let* $u_\varepsilon \to u$ *in* $L^1_{\rm loc}(\mathbf{R})$. *Then for a.e.* $y \in (0,1)$ *we have* $T^\varepsilon_y u_\varepsilon \to u$ *in* $L^1_{\rm loc}(\mathbf{R})$.

**Proof** Let $S > 0$ be fixed. We want to show that for a.e. $y \in (0,1)$ $T^\varepsilon_y u_\varepsilon \to u$ in $L^1(-S, S)$. We have

$$\begin{aligned} \int_0^1\int_{-S}^S |T^\varepsilon_y u_\varepsilon(x) - u(x)|\,dx\,dy &\le \Big(\Big[\frac{1}{\varepsilon}\Big] + 1\Big)\int_0^\varepsilon\int_{-S}^S |T^\varepsilon_y u_\varepsilon(x) - u(x)|\,dx\,dy \\ &\le \frac{1+\varepsilon}{\varepsilon}\int_{-S}^S\int_0^\varepsilon |T^\varepsilon_y u_\varepsilon(x) - u(x)|\,dy\,dx\,. \end{aligned}$$

Fix $x \in [-S, S]$. If $0 < y < x - [x/\varepsilon]\varepsilon$ then $\varepsilon[x/\varepsilon] < x - y < \varepsilon([x/\varepsilon] + 1)$, i.e., $[(x - y)/\varepsilon] = [x/\varepsilon]$, so that

$$T_y^\varepsilon u_\varepsilon(x) = u_\varepsilon\Big(y + \varepsilon\Big[\frac{x}{\varepsilon}\Big]\Big),$$

and

$$\begin{aligned}\int_0^{x-[x/\varepsilon]\varepsilon} |T_y^\varepsilon u_\varepsilon(x) - u(x)|\,dy &= \int_0^{x-[x/\varepsilon]\varepsilon} \Big|u_\varepsilon\Big(y + \varepsilon\Big[\frac{x}{\varepsilon}\Big]\Big) - u(x)\Big|\,dy \\ &= \int_{[x/\varepsilon]\varepsilon}^{x} |u_\varepsilon(t) - u(x)|\,dt\,.\end{aligned}$$

Conversely, if $x - [x/\varepsilon]\varepsilon < y < \varepsilon$ we get $[(x - y)/\varepsilon] = [x/\varepsilon] - 1$,

$$T_y^\varepsilon u_\varepsilon(x) = u_\varepsilon\Big(y + \varepsilon\Big[\frac{x}{\varepsilon}\Big] - \varepsilon\Big),$$

and

$$\int_{x-[x/\varepsilon]\varepsilon}^{\varepsilon} |T_y^\varepsilon u_\varepsilon(x) - u(x)|\,dy = \int_{x-\varepsilon}^{[x/\varepsilon]\varepsilon} |u_\varepsilon(t) - u(x)|\,dt\,.$$

Summing up, we get eventually

$$\int_0^{\varepsilon} |T_y^\varepsilon u_\varepsilon(x) - u(x)|\,dy = \int_{x-\varepsilon}^{x} |u_\varepsilon(t) - u(x)|\,dt\,,$$

and

$$\begin{aligned}\int_0^1 \int_{-S}^{S} |T_y^\varepsilon u_\varepsilon(x) - u(x)|\,dx\,dy &\le \frac{1+\varepsilon}{\varepsilon} \int_{-S}^{S} \int_{x-\varepsilon}^{x} |u_\varepsilon(t) - u(x)|\,dt\,dx \\ &\le (1+\varepsilon) \int_{-S}^{S} \frac{1}{\varepsilon} \int_{x-\varepsilon}^{x} |u_\varepsilon(t) - u(t)|\,dt\,dx \\ &\quad + (1+\varepsilon) \int_{-S}^{S} \frac{1}{\varepsilon} \int_{x-\varepsilon}^{x} |u(t) - u(x)|\,dt\,dx \\ &\le (1+\varepsilon) \int_{-S-\varepsilon}^{S} |u_\varepsilon(t) - u(t)|\,dt \\ &\quad + (1+\varepsilon) \int_{-S}^{S} \frac{1}{\varepsilon} \int_{x-\varepsilon}^{x} |u(t) - u(x)|\,dt\,dx\,.\end{aligned}$$

Both these terms vanish as $\varepsilon \to 0^+$; hence,

$$\lim_{\varepsilon \to 0^+} \int_0^1 \|T_y^\varepsilon u_\varepsilon - u\|_{L^1(-S,S)}\,dy = 0,$$

which implies the thesis. □

**Remark 3.37** If $u_\varepsilon$ is as above, and $v_\varepsilon(x) = T_y^\varepsilon u_\varepsilon(x+\varepsilon)$, then we also have $v_\varepsilon \to u$ in $L^1_{\rm loc}(\mathbf{R})$. As a consequence, for every $\psi$ with values in $[0,1]$, the convex combination $\psi T_y^\varepsilon u_\varepsilon + (1-\psi)v_\varepsilon$ converges to $u$. This a complicated way of saying that for almost every choice of $y$ all functions with values between $u_\varepsilon(y+k\varepsilon)$ and $u_\varepsilon(y+(k+1)\varepsilon)$ on the interval $y+\varepsilon(k,k+1)$ still converge to $u$ in $L^1_{\rm loc}(\mathbf{R})$. In particular we are free to choose, on the interval $y+\varepsilon(k,k+1)$, the affine interpolation between the value at the points $y+\varepsilon k$ and $y+\varepsilon(k+1)$, or the constant value $u_\varepsilon(y+\varepsilon k)$, indifferently.

**Proposition 3.38** *In the hypotheses of Theorem* 3.33 *we get*

$$\Gamma\text{-}\liminf_{\varepsilon\to 0+} F_\varepsilon(u,I) \geq F(u,I)$$

*with respect to the convergence in* $L^1_{\rm loc}(\mathbf{R})$ *for all open sets* $I \subset \mathbf{R}$.

**Proof** We begin with the case $I$ a bounded open interval. Let $u_j \to u$ in $L^1_{\rm loc}(\mathbf{R})$. Let

$$g_j(t) = \begin{cases} f_{\varepsilon_j}\Big(\dfrac{(u_j(t+\varepsilon_j)-u_j(t))^2}{\varepsilon_j}\Big) & \text{if } t \in I \\ 0 & \text{if } t \in \mathbf{R}\setminus I, \end{cases}$$

and let

$$\Phi_j(t) = \sum_{k\in\mathbf{Z}} g_j(t+k\varepsilon_j), \qquad t \in (0,\varepsilon_j)$$

extended by periodicity to all $\mathbf{R}$. Note that the sum takes into account indeed only a finite number of indices $k$. We can write

$$\begin{aligned} F_{\varepsilon_j}(u_j,I)) &= \frac{1}{\varepsilon_j}\int_{\mathbf{R}} g_j(t)\,dt + \frac{1}{\varepsilon_j}\sum_{k\in\mathbf{Z}}\int_{k\varepsilon_j}^{(k+1)\varepsilon_j} g_j(t)\,dt \\ &= \frac{1}{\varepsilon_j}\int_0^{\varepsilon_j}\Phi_j(t)\,dt = \int_0^1 \Phi_j(t)\,dt\,. \end{aligned}$$

Up to considering a subsequence, we can suppose that there exists the limit

$$M = \lim_j F_{\varepsilon_j}(u_j,I) < +\infty.$$

With fixed $\eta > 0$. let

$$A_j = \{y \in (0,1):\ \Phi(y) \leq F_{\varepsilon_j}(u_j,I)+\eta\},$$

and $B_j = (0,1)\setminus A_j$. We have

$$|B_j| \leq \frac{1}{F_{\varepsilon_j}(u_j)+\eta}\int_{B_j}\Phi(t)\,dt \leq \frac{F_{\varepsilon_j}(u_j)}{F_{\varepsilon_j}(u_j)+\eta} \quad\to\quad \frac{M}{M+\eta};$$

hence, we can suppose that $|B_j| \le K$, where $K < 1$ does not depend on $j$, so that $|A_j| \ge 1 - K > 0$. Thanks to Lemma 3.36 we can apply Egorov's Theorem to the sequence of functions $\Psi_j(y) = \|T_y^{\varepsilon_j} u_j - u\|_{L^1_{\rm loc}(\mathbf{R})}$ and obtain the existence of a set $D$ with $|D| \le (1-K)/2$ such that $(\Psi_j)$ converge uniformly on $(0,1) \setminus D$. We can then choose for all $j \in \mathbf{N}$ a point $y_j \in A_j \setminus D$, such that

$$T_{y_j}^{\varepsilon_j} u_j \to u \qquad \text{in } L^1_{\rm loc}(\mathbf{R})$$

and

$$F_{\varepsilon_j}(u_j, I) + \eta \ge \Phi_j(y_j)$$

at the same time. We can suppose for the sake of notation that $y_j = 0$ for all $j$. Note that this is not a restriction, up to a translation of $y_j - [y_j/\varepsilon_j]\varepsilon_j$.

The last inequality above implies that

$$F_{\varepsilon_j}(u_j, I) + \eta \ge \sum_{k \in J_j} f_{\varepsilon_j}\Big(\frac{(u_j((k+1)\varepsilon_j) - u_j(k\varepsilon_j))^2}{\varepsilon_j}\Big),$$

where

$$J_j = \{k \in \mathbf{Z} : \ (\varepsilon_j k, \varepsilon_j(k+1)) \subset I\}.$$

We write $J_j = \{k_0^j, k_1^j (= k_0^j + 1), \dots, k_{N_j}^j\}$,

$$J_j^1 = \{k \in J_j : \ (u_j((k+1)\varepsilon_j) - u_j(k\varepsilon_j))^2 \le C\varepsilon_j\}, \qquad J_j^2 = J_j \setminus J_j^1,$$

and define $v_j$ as follows

$$v_j(t) = \begin{cases} \Big(\dfrac{t}{\varepsilon_j} - k\Big) u_j(\varepsilon_j(k+1)) & \\ \qquad + \Big((k+1) - \dfrac{t}{\varepsilon_j}\Big) u_j(k\varepsilon_j) & \text{in } \varepsilon_j(k, k+1),\ k \in J_j^1 \\ u_j(k\varepsilon_j) & \text{in } \varepsilon_j(k, k+1),\ k \in J_j^2 \\ u_j(k_0^j \varepsilon_j) & \text{if } t \le k_0^j \varepsilon_j \\ u_j((k_{N_j}^j + 1)\varepsilon_j) & \text{if } t \ge (k_{N_j}^j + 1)\varepsilon_j. \end{cases}$$

By Remark 3.37 $v_j \to u$ in $L^1(I)$. Moreover, $v_j \in SBV(I)$, and

$$\begin{aligned} &\sum_{k \in J_j} f_{\varepsilon_j}\Big(\frac{(u_j((k+1)\varepsilon_j) - u_j(k\varepsilon_j))^2}{\varepsilon_j}\Big) \\ &= \sum_{k \in J_j^1} \varepsilon_j h\Big(\frac{|(u_j((k+1)\varepsilon_j) - u_j(k\varepsilon_j)|}{\varepsilon_j}\Big) \\ &\qquad + \sum_{k \in J_j^2} \varphi\Big(|(u_j((k+1)\varepsilon_j) - u_j(k\varepsilon_j)|\Big) \end{aligned}$$

$$= \int_I h(|v_j'|)\,dt + \sum_{S(v_j)\cap I} \varphi(|v_j^+ - v_j^-|),$$

so that

$$F_{\varepsilon_j}(u_j, I) \geq F(v_j, I) - \eta.$$

Recalling the compactness and lower semicontinuity theorems in SBV, we obtain that $u \in SBV(I)$, and that

$$\liminf_j F_{\varepsilon_j}(u_j, I) \geq F(u, I) - \eta.$$

The arbitrariness of $\eta > 0$ gives the thesis in the case $I$ bounded open interval.

Clearly the result still holds if $I$ is a finite union of bounded open intervals with disjoint closures, and eventually for arbitrary $I$ by approximation from the interior. □

**Proposition 3.39** *Let $h$ and $\varphi$ satisfy the hypotheses of Theorem 3.33. Let $g_\varepsilon : [0, +\infty) \to [0, +\infty)$ be a sequence of Borel functions satisfying*

(i) *we have*

$$\frac{1}{\varepsilon} g_\varepsilon(\varepsilon z^2) \to h(|z|) \text{ for all } z \in \mathbf{R} \tag{3.83}$$

*uniformly on compact sets;*

(ii) *we have*

$$g_\varepsilon\Big(\frac{z^2}{\varepsilon}\Big) \to \varphi(|z|) \text{ for all } z \in \mathbf{R}. \tag{3.84}$$

*Let $I$ be an open interval. Then we have*

$$\limsup_{\varepsilon \to 0+} \frac{1}{\varepsilon} \int_I g_\varepsilon\Big(\frac{(u(t+\varepsilon) - u(t))^2}{\varepsilon}\Big)\,dt \leq F(u, I)$$

*for all $u \in SBV_{\rm loc}(\mathbf{R})$ with $\#(S(u)) < +\infty$, $\sup I \notin S(u)$, and $u \in C^1(\mathbf{R} \setminus S(u))$.*

**Proof** As usual it is sufficient to deal with the case $\#(S(u)) = 1$. Hence, we may suppose that $S(u) = 0$ and $I = (a, b)$, so that

$$\frac{1}{\varepsilon} \int_I g_\varepsilon\Big(\frac{(u(t+\varepsilon) - u(t))^2}{\varepsilon}\Big)\,dt = \frac{1}{\varepsilon} \int_a^{-\varepsilon} g_\varepsilon\Big(\frac{(u(t+\varepsilon) - u(t))^2}{\varepsilon}\Big)\,dt$$
$$+ \frac{1}{\varepsilon} \int_0^b g_\varepsilon\Big(\frac{(u(t+\varepsilon) - u(t))^2}{\varepsilon}\Big)\,dt + \frac{1}{\varepsilon} \int_{-\varepsilon}^0 g_\varepsilon\Big(\frac{(u(t+\varepsilon) - u(t))^2}{\varepsilon}\Big)\,dt\,.$$

Since we have

$$\frac{(u(t+\varepsilon) - u(t))^2}{\varepsilon^2} \to (u'(t))^2 \text{ if } t \notin (-\varepsilon, 0),$$

and

$$(u(t+\varepsilon) - u(t))^2 \to (u(0+) - u(0-))^2 \text{ if } t \in (-\varepsilon, 0),$$

by (i) and (ii) we then get the required inequality. □

**Proposition 3.40** *Let $g_\varepsilon$ be as in the previous proposition and let*

$$G_\varepsilon(u,I)=\frac{1}{\varepsilon}\int_I g_\varepsilon\Big(\frac{(u(t+\varepsilon)-u(t))^2}{\varepsilon}\Big)\,dt$$

*be defined for $u\in L^1_{\rm loc}(\mathbf{R})$. Then we have*

$$\Gamma\text{-}\limsup_{\varepsilon\to0+} G_\varepsilon(u,I)\le F(u,I)$$

*for all open intervals $I$ and $u\in L^1_{\rm loc}(\mathbf{R})$.*

**Proof** It clearly suffices to consider the case $F(u,I)<+\infty$. If $u$ satisfies the hypotheses of the previous proposition it suffices to choose $u_\varepsilon=u$ as a recovery sequence. If all hypotheses are satisfied, but $b=\sup I\in S(u)$, we choose

$$u_\varepsilon(t)=\begin{cases}u(t) & \text{if } t\notin(b,b+\varepsilon)\\ u(b-) & \text{if } t\in(b,b+\varepsilon).\end{cases}$$

In the general case, we proceed by density, using Proposition 3.27. □

**Proof of Theorem 3.33** The proof is a direct consequence of Proposition 3.38 and of the proposition above, after remarking that $g_\varepsilon=f_\varepsilon$ trivially satisfies the hypotheses. □

From Theorem 3.33 we deduce an approximation of the Mumford-Shah functional.

**Theorem 3.41** *Let $f:[0,+\infty)\to[0,+\infty)$ be a Borel function satisfying*

$$\inf\{f(z):\ z\ge c\}>0 \qquad \textit{for all } c>0\,, \tag{3.85}$$

*and such that the limits*

$$\lim_{z\to0+}\frac{f(z)}{z}=a,\qquad \lim_{z\to+\infty}f(z)=b \tag{3.86}$$

*exist and $a,b>0$, and let $F_\varepsilon$ be defined in (3.81); i.e.,*

$$F_\varepsilon(u,I)=\frac{1}{\varepsilon}\int_I f\Big(\frac{(u(t+\varepsilon)-u(t))^2}{\varepsilon}\Big)\,dt,. \tag{3.87}$$

*Then if $I$ is an open interval the family $F_\varepsilon(\cdot,I)$ $\Gamma$-converges as $\varepsilon\to0^+$ to the Mumford-Shah functional $F(\cdot,I)$, with respect to the $L^1_{\rm loc}(\mathbf{R})$-convergence, where*

$$F(u,I)=\begin{cases}a\displaystyle\int_I|u'|^2\,dt+b\#(S(u)\cap I) & \textit{if } u\in SBV_{\rm loc}(I)\\ +\infty & \textit{otherwise}\end{cases} \tag{3.88}$$

*is defined for $u\in L^1_{\rm loc}(\mathbf{R})$.*

**Proof** Let $(a_i)$, $(b_i)$ be two sequences of strictly positive real numbers such that $\sup a_i = a$, $\sup b_i = b$, and for all $i \in \mathbf{N}$ the function $f_i(t) = \min\{a_i t, b_i\}$ satisfies $f_i \leq f$. By comparison with any of the functionals related to these functions we obtain, after applying Theorem 3.33 with $f_\varepsilon = f_i$ for all $\varepsilon$, that the $\Gamma$-$\liminf_{\varepsilon \to 0^+} F_\varepsilon(u, I)$ is finite only if $F(u, I) < +\infty$. Moreover, in this case

$$\Gamma\text{-}\liminf_{\varepsilon \to 0+} F_\varepsilon(u, I) \geq a_i \int_I |u'|^2\, dt + b_i \#(S(u) \cap I),$$

and, after applying Theorem 1.16 we obtain that

$$\Gamma\text{-}\liminf_{\varepsilon \to 0+} F_\varepsilon(u, I) \geq a \int_I |u'|^2\, dt + b \#(S(u) \cap I),$$

As for the opposite inequality, it suffices to apply Proposition 3.40. □

**Remark 3.42** If $f(t) \leq \min\{at, b\}$ for all $t \geq 0$ (e.g., if $f$ is concave) then we also have $F_\varepsilon(u, \mathbf{R}) \leq F(u, \mathbf{R})$ for all $u \in SBV(\mathbf{R})$.

### 3.5.1 *Exercises*

**Exercise 3.17** Show that Theorem 3.33 still holds if we take, in place of $C$ constant, $C = C_\varepsilon$, under the condition

$$\varepsilon C_\varepsilon \to 0 \qquad \text{and} \qquad C_\varepsilon \varepsilon^{-1} \to +\infty, \tag{3.89}$$

as $\varepsilon \to 0^+$.

**Exercise 3.18** Let $h : \mathbf{R} \to [0, +\infty)$ be convex and let $\varphi : \mathbf{R} \to [0, +\infty)$ be subadditive and lower semicontinuous, satisfying

$$\lim_{t \to \pm\infty} \frac{h(t)}{|t|} = +\infty, \qquad \lim_{s \to 0} \frac{\varphi(s)}{|s|} = +\infty. \tag{3.90}$$

Let $C_\varepsilon$ satisfy (3.89), and define

$$g_\varepsilon(z) = \begin{cases} h(z) & |z| \leq \sqrt{C_\varepsilon / \varepsilon} \\ \dfrac{1}{\varepsilon}\varphi(\varepsilon z) & |z| > \sqrt{C_\varepsilon / \varepsilon}. \end{cases} \tag{3.91}$$

Prove that the functionals $G_\varepsilon$ defined on $L^1_{\rm loc}(\mathbf{R})$ by

$$G_\varepsilon(u) = \int_{-\infty}^{+\infty} g_\varepsilon\Big(\frac{u(x+\varepsilon) - u(x)}{\varepsilon}\Big)\, dx \tag{3.92}$$

$\Gamma$-converge as $\varepsilon \to 0^+$ with respect to the $L^1_{\rm loc}(\mathbf{R})$ convergence to the functional $G$ given by

$$G(u) = \begin{cases} \displaystyle\int_{-\infty}^{+\infty} h(u')\,dt + \sum_{t\in S(u)} \varphi(u(t+) - u(t-)) & u \in SBV_{\rm loc}(\mathbf{R}) \\ +\infty & \text{otherwise} \end{cases} \tag{3.93}$$

on $L^1_{\rm loc}(\mathbf{R})$.

**Exercise 3.19** Prove that the Γ-limit as $\varepsilon \to 0^+$ of

$$F_\varepsilon(u) = \frac{1}{\varepsilon}\int_{-\infty}^{+\infty} \sqrt{|u(x+\varepsilon) - u(x)|}\,dx \qquad u \in L^1_{\rm loc}(\mathbf{R})$$

is the functional $F$ defined by

$$F(u) = \begin{cases} \displaystyle\sum_{S(u)} \sqrt{|u^+ - u^-|} & u \in SBV_{\rm loc}(\mathbf{R}) \text{ and } u' = 0 \text{ a.e.} \\ +\infty & \text{otherwise} \end{cases}$$

on $L^1_{\rm loc}(\mathbf{R})$.

*Hint*: (Γ-liminf) With fixed $k \in \mathbf{N}$, compare $F$ with the Γ-limit of $G_\varepsilon$ defined as in the exercise above, where $h(z) = kz^2$, $\varphi(t) = \sqrt{|t|}$, and $C_\varepsilon = \varepsilon^{1/3}k^{-4/3}$. Note that $F_\varepsilon \geq G_\varepsilon$ for all $\varepsilon > 0$ so that $F \geq G$, where $G$ is given by the previous exercise. By the arbitrariness of $k \in \mathbf{N}$ obtain the lower semicontinuity inequality.

(Γ-limsup) Check that if $u \in SBV_{\rm loc}(\mathbf{R})$ with $u' = 0$ a.e. and $\#(S(u)) < +\infty$ then $\lim_{\varepsilon\to 0^+} F_\varepsilon(u) = \sum_{S(u)} \sqrt{|u^+ - u^-|}$. For a general $u$ proceed by density.

# 4

# A GENERAL APPROACH TO APPROXIMATION

## 4.1 A lower inequality by slicing

In this section we describe a fruitful method to recover the lower semicontinuity inequality for $\Gamma$-limits through the study of one-dimensional problems. We will use one dimensional sections of $GSB$ functions. For this purpose, we restate some of the results illustrated in Section 1.8.1, using the notation introduced therein.

**Theorem 4.1** (a) *Let $u \in GSBV(\Omega)$. Then for all $\xi \in S^{n-1}$ the function $u_{\xi,y}$ belongs to $GSBV(\Omega_{\xi,y})$ for $\mathcal{H}^{n-1}$-a.a. $y \in \Pi_\xi$. Moreover, for such $y$ we have*

$$u'_{\xi,y}(t) = \langle \nabla u(y+t\xi), \xi\rangle \text{ for a.a. } t \in \Omega_{\xi,y} \tag{4.1}$$

$$S(u_{\xi,y}) = \{t \in \mathbf{R} : \ y+t\xi \in S(u)\}, \tag{4.2}$$

$$v(t\pm) = u^{\pm}(y+t\xi) \qquad \text{or} \qquad v(t\pm) = u^{\mp}(y+t\xi), \tag{4.3}$$

*according to the cases $\langle \nu_u, \xi\rangle > 0$ or $\langle \nu_u, \xi\rangle < 0$ (the case $\langle \nu_u, \xi\rangle = 0$ being negligible) and for all Borel functions $g$*

$$\int_{\Pi_\xi} \sum_{t \in S(u_{\xi,y})} g(t)\, d\mathcal{H}^{n-1}(y) = \int_{S(u)} g(x) |\langle \nu_u, \xi\rangle| d\mathcal{H}^{n-1}. \tag{4.4}$$

(b) *Conversely, if $u \in L^1(\Omega)$ and for all $\xi \in \{e_1, \dots, e_n\}$ and for a.a. $y \in \Pi_\xi$ $u_{\xi,y} \in SBV(\Omega_{\xi,y})$ and*

$$\int_{\Pi_\xi} |Du_{\xi,y}|(\Omega_{\xi,y})\, d\mathcal{H}^{n-1}(y) < +\infty, \tag{4.5}$$

*then $u \in SBV(\Omega)$.*

**Remark 4.2** We will apply frequently Part (b) of the Theorem above to functions $u \in L^\infty(\Omega)$ such that an estimate of the form

$$\int_{\Pi_\xi} \Big( \int_{\Omega_{\xi,y}} |u'_{\xi,y}|\, dt + \#(S(u_{\xi,y}))\Big)\, d\mathcal{H}^{n-1}(y) < +\infty$$

holds. Note that this implies (4.5).

### 4.1.1 *The slicing method*

The use we will make of the theorem above to give an estimate from below of the Γ-liminf of a family of functionals $F_\varepsilon$ can be summarized as follows, and it will be clarified by the examples below.

1. We "localize" the functionals $F_\varepsilon$ highlighting their dependence on the set of integration as usual, defining functionals $F_\varepsilon(\cdot, A)$ for all open subsets $A \subset \Omega$;

2. For all $\xi \in S^{n-1}$ and for all $y \in \Pi_\xi$, we find functionals $F_\varepsilon^{\xi,y}(v, I)$, defined for $I \subset \mathbf{R}$ and $v \in L^1(I)$, such that setting

$$F_\varepsilon^\xi(u, A) = \int_{\Pi_\xi} F_\varepsilon^{\xi,y}(u_{\xi,y}, A_{\xi,y})\, d\mathcal{H}^{n-1}(y) \tag{4.6}$$

we have $F_\varepsilon(u, A) \geq F_\varepsilon^\xi(u, A)$. In this step we use Theorem 4.1(a), or sometimes simply Fubini's Theorem;

3. We compute the $\Gamma\text{-}\liminf_{\varepsilon\to 0} F_\varepsilon^{\xi,y}(v, I) = F^{\xi,y}(v, I)$ and define

$$F^\xi(u, A) = \int_{\Pi_\xi} F^{\xi,y}(u_{\xi,y}, A_{\xi,y})\, d\mathcal{H}^{n-1}(y)\,. \tag{4.7}$$

4. From Fatou's Lemma we have, if $u_\varepsilon \to u$,

$$\begin{aligned}\liminf_{\varepsilon\to 0+} F_\varepsilon(u_\varepsilon, A) &\geq \liminf_{\varepsilon\to 0+} F_\varepsilon^\xi(u_\varepsilon, A)\\ &= \liminf_{\varepsilon\to 0+} \int_{\Pi_\xi} F_\varepsilon^{\xi,y}((u_\varepsilon)_{\xi,y}, A_{\xi,y})\, d\mathcal{H}^{n-1}(y)\\ &\geq \int_{\Pi_\xi} \liminf_{\varepsilon\to 0+} F_\varepsilon^{\xi,y}((u_\varepsilon)_{\xi,y}, A_{\xi,y})\, d\mathcal{H}^{n-1}(y)\\ &\geq \int_{\Pi_\xi} F^{\xi,y}(u_{\xi,y}, A_{\xi,y})\, d\mathcal{H}^{n-1}(y)\\ &= F^\xi(u, A)\,.\end{aligned}$$

Hence, we deduce that

$$\Gamma\text{-}\liminf_{\varepsilon\to 0+} F_\varepsilon(u, A) \geq F^\xi(u, A)\,,$$

for all $\xi \in S^{n-1}$;

5. From estimates from below on $F^{\xi,y}$, and Theorem 4.1(b), or Remark 4.2, we deduce that if $u \in L^\infty(\Omega)$ then $\Gamma\text{-}\liminf_{\varepsilon\to 0+} F_\varepsilon(u, A)$ is finite only if $u \in SBV(A)$. By an approximation argument we see that if $u \in L^1(\Omega)$ only this holds if $u \in GSBV(A)$;

6. We prove the existence of Borel functions $f^\xi, g^\xi$ such that setting

$$f^{\xi,y}(t, s, z) = f^\xi(y + t\xi, s, z), \qquad g^{\xi,y}(t, v, w) = g^\xi(y + t\xi, v, w) \tag{4.8}$$

we have

$$F^{\xi,y}(v,I) \geq \int_I f^{\xi,y}(t,v,v')\,dt + \sum_{S(v)} g^{\xi,y}(t,v^+,v^-), \tag{4.9}$$

and by Theorem 4.1 we deduce that

$$F^{\xi}(u,A) \geq \int_A f^{\xi}(x,u,\langle\nabla u,\xi\rangle)\,dx + \int_{S(u)\cap A} g^{\xi}(x,u^+,u^-)|\langle\nu_u,\xi\rangle|\,d\mathcal{H}^{n-1} \tag{4.10}$$

if $u \in GSBV(\Omega)$;

7. We check that if $u \in GSBV(\Omega)$ then the set function

$$\mu(A) = \Gamma\text{-}\liminf_{\varepsilon\to 0+} F_\varepsilon(u,A)$$

is superadditive on open sets with disjoint compact closures. Take $u$ such that $\mu(\Omega) < +\infty$. Using Theorem 1.16 we conclude that

$$\Gamma\text{-}\liminf_{\varepsilon\to 0+} F_\varepsilon(u,A) \geq \int_A \tilde{f}(x,u,\nabla u)\,dx + \int_{S(u)\cap A} \tilde{g}(x,u^+,u^-,\nu_u)\,d\mathcal{H}^{n-1} \tag{4.11}$$

if $u \in GSBV(\Omega)$, where

$$\tilde{f}(x,s,z) = \sup_i f^{\xi_i}(x,s,\langle z,\xi_i\rangle),$$

$$\tilde{g}(x,v,w,\nu) = \sup_i g^{\xi_i}(x,v,w)|\langle\nu,\xi_i\rangle|,$$

and $(\xi_i)$ is a fixed sequence in $S^{n-1}$. By varying $(\xi_i)$ we can optimize the estimate.

### 4.1.2 *A lower estimate for the perimeter approximation*

We now give a lower estimate for the $\Gamma$-lower limit of the family of functionals

$$F_\varepsilon(u) = \begin{cases} \dfrac{1}{\varepsilon p'}\displaystyle\int_\Omega W(u)\,dx + \frac{1}{p}\varepsilon^{p-1}\int_\Omega \varphi^p(\nabla u)\,dx & \text{if } u \in W^{1,p}(\Omega) \\ +\infty & \text{otherwise,} \end{cases} \tag{4.12}$$

defined on $L^1(\Omega)$. We suppose that $W$ satisfies the hypotheses of Section 3.2.1, and that $\varphi : \mathbf{R}^n \to [0,+\infty)$ is a norm on $\mathbf{R}^n$. We will prove the following estimate.

**Proposition 4.3** *Under the hypotheses above, we have*

$$\Gamma\text{-}\liminf_{\varepsilon\to 0+} F_\varepsilon(u) \geq P_\varphi(u), \tag{4.13}$$

*where* $P_\varphi : L^1(\Omega) \to [0,+\infty]$ *is the* anisotropic perimeter functional *defined by*

$$P_\varphi(u) = \begin{cases} c_p \int_{S(u)} \varphi(\nu_u) d\mathcal{H}^{n-1} & \textit{if } u \in SBV(\Omega) \textit{ and } u \in \{0,1\} \textit{ a.e.} \\ +\infty & \textit{otherwise,} \end{cases} \tag{4.14}$$

*and* $c_p = \int_0^1 (W(s))^{1/p'}\, ds$.

**Proof** We follow the steps outlined in Section 4.1.1.

1. The localized functionals are

$$F_\varepsilon(u,A) = \begin{cases} \frac{1}{\varepsilon p'} \int_A W(u)\,dx + \frac{1}{p}\varepsilon^{p-1} \int_A \varphi^p(\nabla u)\,dx & \text{if } u \in W^{1,p}(\Omega) \\ +\infty & \text{otherwise,} \end{cases} \tag{4.15}$$

defined on $L^1(\Omega) \times \mathcal{A}(\Omega)$.

2. Let $\varphi_*$ be the *dual norm* of $\varphi$. Note that $\varphi$ is linked to $\varphi_*$ by the following relationships:

$$\varphi(\nu) = \max\Big\{ \frac{1}{\varphi_*(\xi)} \langle \nu, \xi\rangle : \xi \in S^{n-1}\Big\}. \tag{4.16}$$

We then choose

$$F_\varepsilon^{\xi,y}(v,I) = \begin{cases} \frac{1}{\varepsilon p'} \int_I W(v)\,dt + \frac{1}{p}\varepsilon^{p-1} \frac{1}{\varphi_*^p(\xi)} \int_I |v'|^p\,dt & \text{if } u \in W^{1,p}(I) \\ +\infty & \text{otherwise} \end{cases} \tag{4.17}$$

(independent of $y$). We then have, by Fubini's Theorem,

$$F_\varepsilon^{\xi}(u,A) = \begin{cases} \frac{1}{\varepsilon p'} \int_A W(u)\,dx + \frac{1}{p}\varepsilon^{p-1} \frac{1}{\varphi_*^p(\xi)} \int_I |\langle \xi, Du\rangle|^p\,dx & \text{if } \langle \xi, Du\rangle << \mathcal{L}_n \\ +\infty & \text{otherwise.} \end{cases} \tag{4.18}$$

Note that (4.16) implies that $F_\varepsilon^\xi \le F_\varepsilon$.

3. By Theorem 3.10 the $\Gamma$-limit

$$F^{\xi,y}(v,I) = \Gamma\text{-}\lim_{\varepsilon\to 0} F_\varepsilon^{\xi,y}(v,I) \tag{4.19}$$

exists and, since we can write

$$F_\varepsilon^{\xi,y}(v,I) = \frac{1}{\varphi_*^p(\xi)} \Big( \frac{1}{\varepsilon p'} \int_I \varphi_*^p(\xi) W(v)\,dt + \frac{1}{p}\varepsilon^{p-1} \int_I |v'|^p\,dt \Big) \tag{4.20}$$

if $v \in W^{1,p}(I)$, we have

$$F^{\xi,y}(v,I) = \begin{cases} \frac{1}{\varphi_*(\xi)} c_p \#(S(v)) & \text{if } v \in \{0,1\} \text{ a.e. on } I, \\ +\infty & \text{otherwise.} \end{cases} \tag{4.21}$$

We define $F^{\xi}$ as in (4.7). Note that $F^{\xi}(u,A)$ is finite if and only if $u \in \{0,1\}$ a.e. in $\Omega$, $u_{\xi,y} \in SBV(A_{\xi,y})$ for $\mathcal{H}^{n-1}$-a.a. $y \in \Pi_{\xi}$, and

$$\int_{\Pi_\xi} \#(S(u_{\xi,y}))d\mathcal{H}^{n-1}(y) < +\infty . \tag{4.22}$$

We then get

$$\int_{\Pi_\xi} |Du_{\xi,y}|(\Omega_{\xi,y})\, d\mathcal{H}^{n-1}(y) \le \int_{\Pi_\xi} 2\#(S(u_{\xi,y}))\, d\mathcal{H}^{n-1}(y) < +\infty . \tag{4.23}$$

4. From Fatou's Lemma we deduce that

$$\Gamma\text{-}\liminf_{\varepsilon\to 0+} F_\varepsilon(u,A) \ge F^{\xi}(u,A),$$

for all $\xi \in S^{n-1}$;

5. By (4.23) and Theorem 4.1(b) we deduce that the $\Gamma$-lower limit $F'(u,A) = \Gamma\text{-}\liminf_{\varepsilon\to 0+} F_\varepsilon(u,A)$ is finite only if $u \in SBV(A)$ and $u \in \{0,1\}$ a.e. Moreover, $F'(u,A) \ge c\mathcal{H}^{n-1}(S(u))$.

6. If $u \in SBV(A)$ and $u \in \{0,1\}$ a.e., from Theorem 4.1(a) we have

$$F^{\xi}(u,A) = c_p \int_{A\cap S(u)} \frac{1}{\varphi_*(\xi)} |\langle \xi, \nu_u\rangle| d\mathcal{H}^{n-1}(y) . \tag{4.24}$$

Hence

$$F'(u,A) \ge c_p \int_{A\cap S(u)} \frac{1}{\varphi_*(\xi)} |\langle \xi, \nu_u\rangle| d\mathcal{H}^{n-1}(y) . \tag{4.25}$$

7. Since all $F_\varepsilon$ are local, then if $u \in SBV(\Omega)$ with $u \in \{0,1\}$ a.e. the set function $\mu(A) = F'(u,A)$ is superadditive on disjoint open sets. From Theorem 1.16 applied with $\lambda = \mathcal{H}^{n-1} \llcorner S(u)$, and $\psi_i(x) = \chi_{S(u)}\varphi_*^{-1}(\xi_i)|\langle \xi_i, \nu_u\rangle|$, where $(\xi_i)$ is a dense sequence in $S^{n-1}$, we conclude that

$$F'(u,A) \ge c_p \int_{S(u)\cap A} \tilde g(\nu_u) d\mathcal{H}^{n-1}, \tag{4.26}$$

where

$$\tilde g(\nu) = \sup_i \Big\{ \frac{1}{\varphi_*(\xi_i)} |\langle \xi_i, \nu\rangle| \Big\} .$$

The thesis follows noticing that $\tilde g = \varphi$ by (4.16). □

**Remark 4.4** We may take take $\varphi(\xi) = |\xi|$. In this case the functional $P = P_\varphi$ reduces to the usual perimeter

$$P(u) = \begin{cases} c_p\,\mathcal{H}^{n-1}(S(u)) & \text{if } u \in SBV(\Omega) \text{ and } u \in \{0,1\} \text{ a.e.} \\ +\infty & \text{otherwise} . \end{cases} \tag{4.27}$$

### 4.1.3 *A lower estimate for the approximation of the Mumford-Shah functionals by elliptic functionals*

We now give a lower estimate for the Γ-lower limit of the family of functionals

$$G_\varepsilon(u,v) = \begin{cases} \displaystyle\int_\Omega \Big(\psi(v)|\nabla u|^2 + \frac{1}{\varepsilon}V(v) + \varepsilon|\nabla v|^2\Big)\,dx & \text{if } u,v \in W^{1,2}(\Omega) \\ & \text{and } 0 \le v \le 1 \\ +\infty & \text{otherwise,} \end{cases} \tag{4.28}$$

defined on $L^1(\Omega)$. We suppose that $\psi$ and $V$ satisfy the hypotheses of Section 3.2.3.

We will prove the following estimate.

**Proposition 4.5** *Under the hypotheses above, we have*

$$\Gamma\text{-}\liminf_{\varepsilon\to 0+} G_\varepsilon(u,v) \ge G(u,v)\,, \tag{4.29}$$

*on* $L^1(\Omega)\times L^1(\Omega)$, *where* $G : L^1(\Omega)\times L^1(\Omega) \to [0,+\infty]$ *is the* Mumford-Shah functional *defined by*

$$G(u,v) = \begin{cases} \displaystyle\int_\Omega |\nabla u|^2\,dx + 4c_V\mathcal{H}^{n-1}(S(u)) & \text{if } u \in GSBV(\Omega) \text{ and } v = 1 \text{ a.e.} \\ +\infty & \text{otherwise,} \end{cases} \tag{4.30}$$

*and* $c_V = \int_0^1 \sqrt{V(s)}\,ds$.

**Proof** We follow the steps outlined in Section 4.1.1.

1. The localized functionals are

$$G_\varepsilon(u,v,A) = \begin{cases} \displaystyle\int_A \Big(\psi(v)|\nabla u|^2 + \frac{1}{\varepsilon}V(v) + \varepsilon|\nabla v|^2\Big)\,dx & \text{if } u,v \in W^{1,2}(\Omega) \\ & \text{and } 0 \le v \le 1 \\ +\infty & \text{otherwise,} \end{cases} \tag{4.31}$$

defined on $L^1(\Omega)\times L^1(\Omega)\times\mathcal{A}(\Omega)$.

2. We simply choose

$$G_\varepsilon^{\xi,y}(u,v,I) = \begin{cases} \displaystyle\int_I \Big(\psi(v)|u'|^2 + \frac{1}{\varepsilon}V(v) + \varepsilon|v'|^2\Big)\,dt & \text{if } u,v \in W^{1,2}(I) \\ & \text{and } 0 \le v \le 1 \\ +\infty & \text{otherwise} \end{cases} \tag{4.32}$$

(independent of $y$) if $u, v \in L^1(\Omega)$ and $I \subset \mathbf{R}$ open and bounded. We then have, by Fubini's Theorem,

$$G_\varepsilon^\xi(u,v,A)=\begin{cases}\displaystyle\int_A\Big(\psi(v)|\langle\nabla u,\xi\rangle|^2+\frac{1}{\varepsilon}V(v)+\varepsilon|\langle\nabla v,\xi\rangle|^2\Big)\,dx\\ \qquad\qquad \text{if } \langle\xi,Du\rangle<<\mathcal{L}_n,\ \langle\xi,Dv\rangle<<\mathcal{L}_n,\ 0\le v\le 1\\ +\infty \qquad\qquad \text{otherwise.}\end{cases} \tag{4.33}$$

3. By Theorem 3.15 the $\Gamma$-limit

$$G^{\xi,y}(u,v,I)=\Gamma\text{-}\lim_{\varepsilon\to 0}G_\varepsilon^{\xi,y}(u,v,I) \tag{4.34}$$

exists and we have

$$G^{\xi,y}(u,v,I)=\begin{cases}\displaystyle\int_I|u'|^2\,dt+4c_V\#(S(v)\cap I) & \text{if } u\in SBV(I)\\ & \text{and } v=1 \text{ a.e. on } I,\\ +\infty & \text{otherwise.}\end{cases} \tag{4.35}$$

We define $G^\xi$ analogously to (4.7). Note that $G^\xi(u,v,A)$ is finite if and only if $v=1$ a.e. in $A$, $u_{\xi,y}\in SBV(A_{\xi,y})$ for $\mathcal{H}^{n-1}$-a.a. $y\in\Pi_\xi$, and

$$\int_{\Pi_\xi}\Big(\int_{A_{\xi,y}}|u'_{\xi,y}|^2\,dt+\#(S(u_{\xi,y})\cap A_{\xi,y})\Big)\,d\mathcal{H}^{n-1}(y)<+\infty\,. \tag{4.36}$$

If in addition $u\in L^\infty(\Omega)$, we get

$$\int_{\Pi_\xi}|Du_{\xi,y}|(\Omega_{\xi,y})\,d\mathcal{H}^{n-1}(y) \tag{4.37}$$

$$\le\int_{\Pi_\xi}\Big(c\Big(\int_{\Omega_{\xi,y}}|u'_{\xi,y}|^2\,dt\Big)^{1/2}+2\|u\|_\infty\#(S(u_{\xi,y}))\Big)\,d\mathcal{H}^{n-1}(y)<+\infty\,.$$

4. From Fatou's Lemma we deduce that

$$\Gamma\text{-}\liminf_{\varepsilon\to 0+}G_\varepsilon(u,v,A)\ge G^\xi(u,v,A)\,,$$

for all $\xi\in S^{n-1}$;

5. If $u\in L^\infty(\Omega)$, by (4.37) and Theorem 4.1(b) we deduce that the $\Gamma$-lower limit $G'(u,v,A)=\Gamma\text{-}\liminf_{\varepsilon\to0+}G_\varepsilon(u,v,A)$ is finite only if $u\in SBV(A)$ and $v=1$ a.e. By a truncation argument it can be easily seen that $G'(u,v,A)\ge G'(u_T,v,A)$ if $u_T=(-T)\vee(T\wedge u)$. Hence, we deduce that $G'(u,v,A)$ is finite only if $v=1$ a.e. and $u\in GSBV(A)$.

6. If $u\in GSBV(A)$ and $v=1$ a.e., from Theorem 4.1(a) we have

$$G^\xi(u,v,A)=\int_A|\langle\nabla u,\xi\rangle|^2\,dx+4c_V\int_{A\cap S(u)}|\langle\xi,\nu_u\rangle|\,d\mathcal{H}^{n-1} \tag{4.38}$$

Hence

$$G'(u,v,A) \geq \int_A |\langle \nabla u, \xi\rangle|^2\,dx + 4c_V \int_{A\cap S(u)} |\langle \xi, \nu_u\rangle|\,d\mathcal{H}^{n-1}. \tag{4.39}$$

7. Since all $G_\varepsilon$ are local, we have that if $u \in GSBV(\Omega)$ and $v = 1$ a.e. then the set function $\mu(A) = G'(u,v,A)$ is superadditive on disjoint open sets. From Theorem 1.16 applied with $\lambda = \mathcal{L}_n + \mathcal{H}^{n-1} \llcorner S(u)$, and

$$\psi_i(x) = \begin{cases} |\langle \nabla u, \xi_i\rangle|^2 & \text{if } x \notin S(u) \\ \\ 4c_V |\langle \xi_i, \nu_u\rangle| & \text{if } x \in S(u), \end{cases}$$

where $(\xi_i)$ is a dense sequence in $S^{n-1}$, we conclude that

$$G'(u,v,A) \geq \int_A |\nabla u|^2\,dx + 4c_V \mathcal{H}^{n-1}(S(u)\cap A), \tag{4.40}$$

and the thesis. □

### 4.1.4 *A lower estimate for the approximation by high-order perturbations*

We now study the Γ-lower limit of the family of functionals

$$F_\varepsilon(u) = \begin{cases} \dfrac{1}{\varepsilon}\displaystyle\int_\Omega f(\varepsilon|\nabla u|^2)\,dx + \varepsilon^3 \int_\Omega \|\nabla^2 u\|^2\,dx & \text{if } u \in W^{2,2}(\Omega) \\ +\infty & \text{otherwise,} \end{cases} \tag{4.41}$$

defined on $L^1(\Omega)$, where $f$ satisfies the hypotheses of Section 3.3.1, for some $\alpha$ and $\beta$. We use denote by $\nabla^2 u$ the Hessian matrix of $u$, and

$$\|A\| = \max\{\langle A\xi, \xi\rangle : \ |\xi| = 1\}.$$

We define $F : L^1(\Omega) \to [0, +\infty]$ by

$$F(u) = \begin{cases} \alpha \displaystyle\int_\Omega |\nabla u|^2\,dx + m(\beta)\int_{S(u)} \sqrt{|u^+ - u^-|}\,d\mathcal{H}^{n-1} & \text{if } u \in GSBV(\Omega) \\ +\infty & \text{otherwise,} \end{cases} \tag{4.42}$$

where $m(\beta) = \beta^{3/4}\left(2\sqrt{3/2} + \sqrt{2/3}\right)$. We will prove the following estimate.

**Proposition 4.6** *Under the hypotheses above, we have*

$$\Gamma\text{-}\liminf_{\varepsilon\to 0^+} F_\varepsilon(u) \geq F(u), \tag{4.43}$$

*on* $L^\infty(\Omega)$.

**Proof** We follow the steps outlined in Section 4.1.1.

1. The localized functionals are

$$F_\varepsilon(u,A)=\begin{cases}\dfrac{1}{\varepsilon}\displaystyle\int_A f(\varepsilon|\nabla u|^2)\,dx+\varepsilon^3\int_A\|\nabla^2 u\|^2\,dx & \text{if } u\in W^{2,2}(\Omega)\\ +\infty & \text{otherwise,}\end{cases}\tag{4.44}$$

defined on $L^1(\Omega)\times\mathcal{A}(\Omega)$.

2. We simply choose

$$F_\varepsilon^{\xi,y}(u,I)=\begin{cases}\dfrac{1}{\varepsilon}\displaystyle\int_I f(\varepsilon|u'|^2)\,dt+\varepsilon^3\int_I|u''|^2\,dt & \text{if } u\in W^{2,2}(I)\\ +\infty & \text{otherwise}\end{cases}\tag{4.45}$$

(independent of $y$) if $u\in L^1(I)$ and $I\subset\mathbf{R}$ open and bounded. We then have, by Fubini's Theorem,

$$F_\varepsilon^{\xi}(u,A)=\begin{cases}\dfrac{1}{\varepsilon}\displaystyle\int_A f(\varepsilon|\langle\nabla u,\xi\rangle|^2)\,dx+\varepsilon^3\int_A|\langle(\nabla^2 u)\xi,\xi\rangle|^2\,dx\\ \qquad\qquad\text{if } u_{\xi,y}\in W^{2,2}(A_{\xi,y}) \text{ for all } y\in\Pi_\xi\\ +\infty \qquad\qquad \text{otherwise.}\end{cases}\tag{4.46}$$

3. By Theorem 3.19 the $\Gamma$-limit

$$F^{\xi,y}(u,I)=\Gamma\text{-}\lim_{\varepsilon\to0}F_\varepsilon^{\xi,y}(u,I)\tag{4.47}$$

exists and we have

$$F^{\xi,y}(u,I)=\begin{cases}\alpha\displaystyle\int_I|u'|^2\,dt+m(\beta)\sum_{I\cap S(u)}\sqrt{|u^+-u^-|} & \text{if } u\in SBV(I)\\ +\infty & \text{otherwise.}\end{cases}\tag{4.48}$$

We define $F^\xi$ as in (4.7). Note that $F^\xi(u,A)$ is finite if and only if $u_{\xi,y}\in SBV(A_{\xi,y})$ for $\mathcal{H}^{n-1}$-a.a. $y\in\Pi_\xi$, and

$$\int_{\Pi_\xi}\Big(\int_{A_{\xi,y}}|u'_{\xi,y}|^2\,dt+\sum_{S(u_{\xi,y})\cap A_{\xi,y}}\sqrt{|u^+_{\xi,y}-u^-_{\xi,y}|}\Big)\,d\mathcal{H}^{n-1}(y)<+\infty.\tag{4.49}$$

If in addition $u\in L^\infty(\Omega)$, we get

$$\int_{\Pi_\xi}|Du_{\xi,y}|(\Omega_{\xi,y})\,d\mathcal{H}^{n-1}(y)\tag{4.50}$$

$$\leq \int_{\Pi_\xi} \Big( c\Big( \int_{\Omega_{\xi,y}} |u'_{\xi,y}|^2\,dt \Big)^{1/2} + \sqrt{2\|u\|_\infty} \sum_{S(u_{\xi,y})} \sqrt{|u^+_{\xi,y} - u^-_{\xi,y}|} \Big)\, d\mathcal{H}^{n-1}(y) < +\infty\,.$$

4. From Fatou's Lemma we deduce that

$$\Gamma\text{-}\liminf_{\varepsilon\to 0+} F_\varepsilon(u, A) \geq F^\xi(u, A)\,,$$

for all $\xi \in S^{n-1}$.

5. If $u \in L^\infty(\Omega)$, by (4.50) and Theorem 4.1(b) we deduce that the $\Gamma$-lower limit $F'(u, A) = \Gamma\text{-}\liminf_{\varepsilon\to 0+} F_\varepsilon(u, A)$ is finite only if $u \in SBV(A)$. Note that in this step the application of a truncation argument to get an estimate in $L^1(A)$ is not trivial.

6. If $u \in GSBV(A)$, from Theorem 4.1(a) we have

$$F^\xi(u, A) = \alpha \int_A |\langle \nabla u, \xi\rangle|^2\,dx + m(\beta) \int_{A\cap S(u)} \sqrt{|u^+ - u^-|}\,|\langle \xi, \nu_u\rangle|\,d\mathcal{H}^{n-1}\,. \tag{4.51}$$

Hence, if $u \in SBV(A) \cap L^\infty(A)$,

$$F'(u, A) \geq \alpha \int_A |\langle \nabla u, \xi\rangle|^2\,dx + m(\beta) \int_{A\cap S(u)} \sqrt{|u^+ - u^-|}\,|\langle \xi, \nu_u\rangle|\,d\mathcal{H}^{n-1}\,. \tag{4.52}$$

7. Since all $F_\varepsilon$ are local, we have that if $u \in SBV(\Omega)\cap L^\infty(\Omega)$ and $F(u) < +\infty$ then the set function $\mu(A) = F'(u, A)$ is superadditive on disjoint open sets. From Theorem 1.16 applied with $\lambda = \mathcal{L}_n + \sqrt{|u^+ - u^-|}\,\mathcal{H}^{n-1} \llcorner S(u)$, and

$$\psi_i(x) = \begin{cases} \alpha |\langle \nabla u, \xi_i\rangle|^2 & \text{if } x \notin S(u) \\ m(\beta)\,|\langle \xi_i, \nu_u\rangle| & \text{if } x \in S(u), \end{cases}$$

where $(\xi_i)$ is a dense sequence in $S^{n-1}$, we conclude that

$$F'(u, A) \geq \int_A |\nabla u|^2\,dx + m(\beta) \int_{A\cap S(u)} \sqrt{|u^+ - u^-|}\,d\mathcal{H}^{n-1}\,, \tag{4.53}$$

and the thesis. □

## 4.2 An upper inequality by density

Given a family of functionals $F_\varepsilon : L^1(\Omega) \to [0, +\infty]$ we use the notation

$$F''(u) := \Gamma\text{-}\limsup_{\varepsilon\to 0+} F_\varepsilon(u)$$

for its $\Gamma$-limsup. While it is usually difficult to prove directly a meaningful upper inequality for $F''$ on the whole $L^1(\Omega)$, a recovery sequence can often be easily constructed if the target function $u$ has some special structure. In order to give an upper estimate of $F''$ by some functional $F : L^1(\Omega) \to [0, +\infty]$ it is therefore useful to proceed as follows.

Step 1. Define a subset $\mathcal{D}$ of $L^1(\Omega)$, dense in $\{F < +\infty\}$ (the domain of $F$), such that for each $u \in L^1(\Omega)$ such that $F(u) < +\infty$ we can find a sequence $(u_j) \subset \mathcal{D}$ such that $u_j \to u$ in $L^1(\Omega)$, and $F(u) = \lim_j F(u_j)$;

Step 2. Prove that we have $F''(u) \le F(u)$ for each $u \in \mathcal{D}$.

By the lower semicontinuity of $F''$ we then conclude that

$$F''(u) \le \liminf_j F''(u_j) \le \lim_j F(u_j) = F(u) \tag{4.54}$$

if $F(u) < +\infty$, so that $F'' \le F$ on $L^1(\Omega)$.

### 4.2.1 *An upper estimate for the perimeter approximation*

We use the same notation as in Section 4.1.2. We want to proceed as outlined above with $F_\varepsilon = P_\varepsilon$ and $F = P_\varphi$. Note that in this case the domain of $F$ is the family of all (characteristic functions of) sets of finite perimeter.

**Proposition 4.7** *Let $\Omega$ be a Lipschitz set. If $E$ is a set of finite perimeter in $\Omega$ then there exists a sequence $(E_j)$ of sets of finite perimeter in $\Omega$, such that*

$$|E \Delta E_j| \to 0, \qquad |D\chi_{E_j}|(\Omega) \to |D\chi_E|(\Omega) \tag{4.55}$$

*as $j \to +\infty$, and for every open set $\Omega'$ with $\Omega \subset\subset \Omega'$ there exist sets $E'_j$ of class $C^\infty$ in $\Omega'$ and such that $E'_j \cap \Omega = E_j$*

**Proof** We can extend $E$ in a neighbourhood of $\partial\Omega$ by a local reflection argument, and then in $\mathbf{R}^n$ to a set of locally finite perimeter which we still denote by $E$.

With fixed $\widetilde{\Omega}$ with $\Omega \subset\subset \widetilde{\Omega}$, we can find a sequence $(\widetilde{E}_j)$ of sets of class $C^\infty$ in $\widetilde{\Omega}$, with $|(E \Delta \widetilde{E}_j) \cap \widetilde{\Omega}| \to 0$ and $|D\chi_{\widetilde{E}_j}|(\widetilde{\Omega}) \to |D\chi_E|(\widetilde{\Omega})$ as $j \to +\infty$ (see Exercise 1.21). Since $|D\chi_E|(\partial\Omega) = 0$ by construction, we also have $|D\chi_{\widetilde{E}_j}|(\Omega) \to |D\chi_E|(\Omega)$ by Proposition 1.36. If $(E_j)$ is any sequence of $C^\infty$ sets of locally finite perimeter coinciding with $(\widetilde{E}_j)$ in a neighbourhood of $\Omega$, then $(E_j)$ satisfies the thesis of the proposition. □

**Remark 4.8** If $E_j$ and $E$ are as in the previous proposition, then for every $\varphi$ positive, convex and positively homogeneous of degree one we have

$$\lim_j \int_{\Omega \cap \partial^* E_j} \varphi(\nu_j)\, d\mathcal{H}^{n-1} = \int_{\Omega \cap \partial^* E} \varphi(\nu)\, d\mathcal{H}^{n-1}, \tag{4.56}$$

where $\nu_j$, $\nu$ denote the respective interior normals. In fact, we can apply Reshetnyak's Theorem 1.40 to the measures

$$\mu_j = \nu_j \mathcal{H}^{n-1} \llcorner \partial^* E_j, \qquad \mu = \nu \mathcal{H}^{n-1} \llcorner \partial^* E,$$

and deduce that $\lim_j \int_\Omega \varphi(\mu_j) = \int_\Omega \varphi(\mu)$, which is equivalent to (4.56).

We define the set $\mathcal{D}$ as the family of all characteristic functions of subsets of $\Omega$ which are restrictions of sets of class $C^\infty$ in a neighbourhood of $\Omega$.

**Remark 4.9** The proposition above shows that $\mathcal{D}$ is dense in $\{F < +\infty\}$, since the latter is the family of all characteristic functions of subsets of finite perimeter in $\Omega$. Moreover, (4.55) shows that all the requirements of Step 1 above are satisfied.

It remains to prove Step 2; i.e., to construct a recovery sequence for characteristic functions of smooth sets.

**Proposition 4.10** *We have $F''(u) \le F(u)$ for each $u \in \mathcal{D}$.*

**Proof** Let $u = \chi_E \in \mathcal{D}$. Since $\partial E$ is of class $C^\infty$ up to the boundary of $\Omega$, for $\eta > 0$ sufficiently small the projection $\pi : \{x \in \Omega : \ \mathrm{dist}\,(x, \partial E) < \eta\} \to \partial E$ is well-defined.

Let $v$ be a minimizer of the problem

$$c_p = \min\Big\{\int_{-\infty}^{+\infty} \Big(\frac{1}{p'}W(u) + \frac{1}{p}|v'|^p\Big)\,dt : \ u(-\infty) = 0, \ u(+\infty) = 1\Big\}.$$

Note that $v_z(t) = v(t/z)$ minimizes the problem

$$z\,c_p = \min\Big\{\int_{-\infty}^{+\infty} \Big(\frac{1}{p'}W(u) + z^p\frac{1}{p}|v'|^p\Big)\,dt : \ u(-\infty) = 0, \ u(+\infty) = 1\Big\}.$$

With fixed $\eta \in (0,1)$, we set $v^\eta = 0 \vee \Big(((1+2\eta)v - \eta) \wedge 1\Big)$. Note that

$$c_p^\eta := \int_{-\infty}^{+\infty} \Big(\frac{1}{p'}W(v^\eta) + \frac{1}{p}|(v^\eta)'|^p\Big)\,dt \longrightarrow c_p, \qquad \text{as } \eta \to 0.$$

We define $d(x) = \mathrm{dist}\,(x, \Omega\setminus E) - \mathrm{dist}\,(x, E)$, the *signed distance function* to $\partial E$, and

$$u_\varepsilon(x) = \begin{cases} v^\eta\Big(\frac{d(x)}{\varepsilon\varphi(\nu(x))}\Big) & \text{if } |d(x)| \le T\varepsilon, \\ 0 & \text{otherwise in } \Omega\setminus E \\ 1 & \text{otherwise in } E, \end{cases}$$

where $T > 0$ is large enough as to have $\mathrm{spt}\,(v^\eta)' \subset [-TM, TM]$, where $M = \max\{\varphi(\nu) : |\nu| \le 1\}$, and $\nu(x) = (x - \pi(x))/|x - \pi(x)|$. Note that this is a good definition if $\varepsilon$ is small enough, as to have a good definition of $\pi$ on $\{|d| \le \varepsilon T\}$, and that $\nu(x)$ coincides with a normal to $E$ at $\pi(x)$.

The function $\tilde d(x) = d(x)/\varphi(\nu(x))$ is Lipschitz on $\{|d| \le T\varepsilon\}$. Moreover, since $1/\varphi$ and $\nu$ are Lipschitz, we have

$$|\nabla u_\varepsilon(x)| \le |\nabla v^\eta(\tilde d(x)/\varepsilon)|\Big(\frac{1}{\varepsilon\varphi(\nu(x))} + c\Big)$$

on $\{|d| \leq T\varepsilon\}$.

If $\Omega'$ is any open set with $\Omega \subset\subset \Omega'$, we now can estimate

$$\begin{aligned}
&\int_\Omega \Big(\frac{1}{\varepsilon p'} W(u_\varepsilon) + \frac{\varepsilon^{p-1}}{p}|\nabla u_\varepsilon|^p\Big)\, dx \\
&\leq (1+c\varepsilon)\int_\Omega \Big(\frac{1}{\varepsilon p'} W(v^\eta(\tilde d(x)/\varepsilon)) + \frac{1}{\varepsilon p \varphi^p(\nu(x))}|\nabla v^\eta(\tilde d(x)/\varepsilon)|^p\Big)\, dx \\
&\leq (1+c\varepsilon)\int_{-T\varepsilon}^{T\varepsilon}\int_{\{d(x)=t\}} \Big(\frac{1}{\varepsilon p'} W(v^\eta(\frac{t}{\varepsilon\varphi(\nu(x))})) \\
&\qquad + \frac{1}{\varepsilon p \varphi^p(\nu(x))}|\nabla v^\eta(\frac{t}{\varepsilon\varphi(\nu(x))})|^p\Big)\, d\mathcal{H}^{n-1}(x)\, dt \\
&\leq (1+c\varepsilon)\int_{\partial E\cap\Omega'}\int_{-\varepsilon T\varphi(\nu_E)}^{-\varepsilon T\varphi(\nu_E)} \Big(\frac{1}{\varepsilon p'} W(v^\eta(t)) \\
&\qquad + \frac{1}{\varepsilon p \varphi^p(\nu_E(x))}|\nabla v^\eta(t)|^p\Big)\, dt\, d\mathcal{H}^{n-1}(y) + o(1) \\
&\leq (1+c\varepsilon)\int_{\partial E\cap\Omega'} c_p \varphi(\nu_E(y))\, d\mathcal{H}^{n-1}(y) + o(1),
\end{aligned}$$

where $y = x + t\nu$; we have used the coarea formula and the fact that $|\nabla d| = 1$ a.e. By the arbitrariness of $\Omega'$ we get

$$\limsup_{\varepsilon\to 0+} F_\varepsilon(u_\varepsilon) \leq c_p \int_{\partial E\cap\Omega} \varphi(\nu_E)\, d\mathcal{H}^{n-1},$$

as desired. □

### 4.2.2 *A density result in SBV*

We recall that $\mathcal{M}^{n-1}(E)$ is the $n-1$ dimensional Minkowski content of $E$, defined by

$$\mathcal{M}^{n-1}(E) = \lim_{\varepsilon\to 0}\frac{1}{2\varepsilon}|\{x \in \mathbf{R}^n : \operatorname{dist}(x, E) < \varepsilon\}|, \tag{4.57}$$

whenever the limit in (4.57) exists.

**Lemma 4.11** *Let $\Omega$ be a bounded open set with Lipschitz boundary. Let $u \in SBV(\Omega)\cap L^\infty(\Omega)$ with $\int_\Omega |\nabla u|^2\, dx + \mathcal{H}^{n-1}(S(u)) < +\infty$ and let $\Omega'$ be a bounded open subset of $\mathbf{R}^n$ such that $\Omega \subset\subset \Omega'$. Then $u$ has an extension $z \in SBV(\Omega')\cap L^\infty(\Omega')$ such that $\int_{\Omega'} |\nabla z|^2\, dx + \mathcal{H}^{n-1}(S(z)) < +\infty$, $\mathcal{H}^{n-1}(S(z)\cap\partial\Omega) = 0$, and $\|z\|_{L^\infty(\Omega')} = \|u\|_{L^\infty(\Omega)}$. Moreover, there exists a sequence $(z_k)$ in $SBV(\Omega')\cap L^\infty(\Omega')$ such that $(z_k)$ converges to $z$ in $L^1(\Omega')$, $(\nabla z_k)$ converges to $\nabla z$ in $L^2(\Omega';\mathbf{R}^n)$, $\|z_k\|_{L^\infty(\Omega')} \leq \|u\|_{L^\infty(\Omega)}$,*

$$\lim\nolimits_k \mathcal{H}^{n-1}(S(z_k)) = \mathcal{H}^{n-1}(S(z)), \tag{4.58}$$

$$\lim_k \mathcal{H}^{n-1}(S(z_k)\cap\overline{\Omega}) = \mathcal{H}^{n-1}(S(z)\cap\Omega) = \mathcal{H}^{n-1}(S(u)), \tag{4.59}$$

$\mathcal{H}^{n-1}(\overline{S}(z_k)\setminus S(z_k)) = 0$, *and* $\mathcal{H}^{n-1}(S(z_k)\cap K) = \mathcal{M}^{n-1}(S(z_k)\cap K)$ *for every compact set* $K \subseteq \Omega'$.

**Proof** Since $\partial\Omega$ is Lipschitz, an extension $z$ of $u$ with the required properties can be easily obtained by a local reflection argument. For every integer $k > 0$ let $z_k$ be a solution to the minimum problem

$$m_k = \min_{v\in SBV(\Omega')}\left(\int_{\Omega'}|\nabla v|^2\,dx + \mathcal{H}^{n-1}(S(v)) + k\int_{\Omega'}|v-z|^p\,dx\right).$$

By an easy truncation argument we have $\|z_k\|_{L^\infty(\Omega')} \le \|z\|_{L^\infty(\Omega')} = \|u\|_{L^\infty(\Omega)}$. Since

$$m_k \le \int_{\Omega'}|\nabla z|^2\,dx + \mathcal{H}^{n-1}(S(z)) < +\infty, \tag{4.60}$$

we immediately have that $(z_k)$ converges to $z$ in $L^1(\Omega')$. By Theorem 2.3 we also have that $(\nabla z_k)$ converges to $\nabla z$ weakly in $L^2(\Omega';\mathbf{R}^n)$ and $\mathcal{H}^{n-1}(S(z)) \le \liminf_k \mathcal{H}^{n-1}(S(z_k))$. These facts, together with (4.60), imply that

$$\begin{aligned}&\int_{\Omega'}|\nabla z|^2\,dx + \mathcal{H}^{n-1}(S(z))\\ &= \lim_k\left(\int_{\Omega'}|\nabla z_k|^2\,dx + \mathcal{H}^{n-1}(S(z_k)) + k\int_{\Omega'}|z_k-z|^p\,dx\right),\end{aligned}$$

so that indeed $(\nabla z_k)$ converges to $\nabla z$ strongly in $L^2(\Omega';\mathbf{R}^n)$ and (4.58) holds. By Theorem 2.3 we also have

$$\begin{aligned}\mathcal{H}^{n-1}(S(z)\cap\Omega) &\le \liminf_k\mathcal{H}^{n-1}(S(z_k)\cap\Omega) \le \liminf_k\mathcal{H}^{n-1}(S(z_k)\cap\overline{\Omega}),\\ \mathcal{H}^{n-1}(S(z)\setminus\Omega) &= \mathcal{H}^{n-1}(S(z)\setminus\overline{\Omega}) \le \liminf_k\mathcal{H}^{n-1}(S(z_k)\setminus\overline{\Omega}),\end{aligned}$$

which, together with (4.58), imply (4.59). We conclude the proof by applying Remark 2.22. □

### 4.2.3 *An upper estimate for the approximation of the Mumford-Shah functionals by elliptic functionals*

We briefly remark how an upper estimate can be recovered for the Γ-limsup of the family of functionals $G_\varepsilon$ defined in (4.28), proceeding as in Section 4.2.1 and using Lemma 4.11.

We use the notation

$$F''(u) = \Gamma\text{-}\limsup_{\varepsilon\to 0+} G_\varepsilon(u,1),$$

and $F(u) = G(u,1)$, where $G$ is given by (4.29), defined for $u \in GSBV(\Omega)$.

**Proposition 4.12** *Let* $\Omega$ *have a Lipschitz boundary. Then we have* $F''(u) \le F(u)$ *for all* $u \in GSBV(\Omega)$.

**Proof** Step 1. Let $\Omega \subset\subset \Omega'$, and let $u \in SBV(\Omega) \cap L^\infty(\Omega)$ with $\nabla u \in L^2(\Omega; \mathbf{R}^n)$, $\mathcal{H}^{n-1}(\overline{S}(u) \setminus S(u)) = 0$, and $\mathcal{H}^{n-1}(S(u) \cap K) = \cap M^{n-1}(S(u) \cap K)$ for every compact set $K \subseteq \Omega'$. Then a recovery sequence can be constructed as in the last part of the proof of Theorem 3.15, letting $v_\varepsilon(x)$ depend on $\mathrm{dist}\,(x, S(u))$.

Step 2. If $u \in SBV(\Omega) \cap L^\infty(\Omega)$ it suffices to apply Lemma 4.11.

Step 3. If $u \in GSBV(\Omega)$ an easy truncation argument gives the thesis. □

Note that the previous proposition also implies that $\Gamma\text{-}\limsup_{\varepsilon\to 0+} G_\varepsilon(u,v) \leq G(u,v)$ on the whole $L^1(\Omega) \times L^1(\Omega)$.

## 4.3 Convergence results

Collecting the lower and upper estimates for the approximation of the perimeter functional (Sections 4.1.2 and 4.2.1) and for the elliptic approximation of the Mumford-Shah functional (Sections 4.1.3 and 4.2.3) we obtain two $\Gamma$-convergence results.

**Theorem 4.13** *Let $\Omega$ be a bounded open set with Lipschitz boundary. Let $p > 1$, let $W : \mathbf{R} \to [0, +\infty)$ be a continuous function such that $W(z) = 0$ if and only if $z \in \{0, 1\}$, and let $\varphi : \mathbf{R}^n \to [0, +\infty)$ be a norm on $\mathbf{R}^n$. Let $F_\varepsilon : L^1(\Omega) \to [0, +\infty]$ be defined by*

$$F_\varepsilon(u) = \begin{cases} \dfrac{1}{\varepsilon p'} \displaystyle\int_\Omega W(u)\,dx + \frac{1}{p}\varepsilon^{p-1} \int_\Omega \varphi^p(\nabla u)\,dx & \text{if } u \in W^{1,p}(\Omega) \\ +\infty & \text{otherwise,} \end{cases} \tag{4.61}$$

*and let $P_\varphi : L^1(\Omega) \to [0, +\infty]$ be defined by*

$$P_\varphi(u) = \begin{cases} c_p \displaystyle\int_{S(u)} \varphi(\nu_u) d\mathcal{H}^{n-1} & \text{if } u \in SBV(\Omega) \text{ and } u \in \{0,1\} \text{ a.e.} \\ +\infty & \text{otherwise,} \end{cases} \tag{4.62}$$

*where $c_p = \int_0^1 \big(W(s)\big)^{1/p'} ds$. Then $\Gamma\text{-}\lim_{\varepsilon\to 0+} F_\varepsilon(u) = P_\varphi(u)$.*

**Proof** The proof is obtained by applying Propositions 4.3 and 4.10. □

**Theorem 4.14** *Let $\Omega$ be a bounded open set with Lipschitz boundary. Let $V : \mathbf{R} \to [0, +\infty)$ be a continuous function such that $V(z) = 0$ if and only if $z = 1$, let $\psi : [0,1] \to [0, +\infty)$ be a lower semicontinuous and increasing function, with $\psi(z) = 0$ if and only if $z = 0$ and $\psi(1) = 1$. Let $G_\varepsilon : L^1(\Omega) \times L^1(\Omega) \to [0, +\infty]$ be defined by*

$$G_\varepsilon(u,v) = \begin{cases} \displaystyle\int_\Omega \Big(\psi(v)|\nabla u|^2 + \frac{1}{\varepsilon} V(v) + \varepsilon |\nabla v|^2\Big)\,dx & \text{if } u, v \in W^{1,2}(\Omega) \\ & \text{and } 0 \leq v \leq 1 \\ +\infty & \text{otherwise,} \end{cases} \tag{4.63}$$

*and let* $G : L^1(\Omega) \times L^1(\Omega) \to [0, +\infty]$ *be defined by*

$$G(u,v) = \begin{cases} \displaystyle\int_\Omega |\nabla u|^2\,dx + 4c_V \mathcal{H}^{n-1}(S(u)) & \text{if } u \in GSBV(\Omega) \text{ and } v = 1 \text{ a.e.} \\ +\infty & \text{otherwise,} \end{cases} \tag{4.64}$$

*where* $c_V = \int_0^1 \sqrt{V(s)}\,ds$. *Then we have* $\Gamma\text{-}\lim_{\varepsilon\to 0^+} G_\varepsilon(u,v) = G(u,v)$. *Moreover, for every choice of strictly positive numbers* $k(\varepsilon) = o(\varepsilon)$, $p \geq 1$ *and* $C > 0$, *and for every* $g \in L^\infty(\Omega)$ *there exists a solution* $(u_\varepsilon, v_\varepsilon)$ *for the minimum problem*

$$m_\varepsilon = \min_{u,v}\Big\{G_\varepsilon(u,v) + k(\varepsilon)\int_\Omega |\nabla u|^2\,dx + C\int_\Omega |u-g|^p\,dx\Big\} \tag{4.65}$$

*with* $\|u_\varepsilon\|_\infty \leq \|g\|_\infty$, *up to subsequences* $u_\varepsilon$ *converge in* $L^1(\Omega)$ *to a solution* $u$ *for the minimum problem*

$$m = \min\Big\{\int_\Omega |\nabla u|^2\,dx + 4c_V\mathcal{H}^{n-1}(S(u)) + C\int_\Omega |u-g|^p\,dx : u \in SBV(\Omega)\Big\}, \tag{4.66}$$

*and* $m_\varepsilon \to m$ *as* $\varepsilon \to 0^+$

**Proof** The proof of the Γ-convergence is contained in Propositions 4.5 and 4.12. To prove the convergence result for $u_\varepsilon$ and $m_\varepsilon$ it suffices to note that by truncation we can always assume that functions $u$ satisfy $\|u\|_\infty \leq \|g\|_\infty$, and repeat the reasoning of Exercises 3.4 and 3.5. Note that from the first part of the proof of Theorem 3.15 and Theorem 4.1 it can be easily shown that $(u_\varepsilon)$ is a precompact family in $L^1(\Omega)$. □

# 5

# NON-LOCAL APPROXIMATION

In this chapter we prove two non-local approximation results for the Mumford-Shah functional. The methods we use draw largely from the ones described in Chapter 4, but must be completed by non-trivial technical arguments.

## 5.1 Non-local approximation of the Mumford-Shah functional

Let $\Omega$ be a bounded open subset of $\mathbf{R}^n$ with Lipschitz boundary and let $f : [0,+\infty) \to [0,+\infty)$ be a non-decreasing lower semicontinuous function such that $a, b > 0$ exist with

$$\lim_{t\to 0+} \frac{f(t)}{t} = a \qquad \text{and} \qquad \lim_{t\to+\infty} f(t) = b\,. \tag{5.1}$$

For every $\varepsilon > 0$ we consider the functionals $F_\varepsilon : L^1(\Omega) \to [0,+\infty]$ defined by

$$F_\varepsilon(u) = \begin{cases} \dfrac{1}{\varepsilon}\displaystyle\int_\Omega f\Big(\varepsilon \fint_{B_\varepsilon(x)\cap\Omega} |\nabla u(y)|^2\, dy\Big)\, dx & \text{if } u \in H^1(\Omega), \\ +\infty & \text{otherwise,} \end{cases} \tag{5.2}$$

and $F : L^1(\Omega) \to [0,+\infty]$ defined by

$$F(u) = \begin{cases} a\displaystyle\int_\Omega |\nabla u(x)|^2\, dx + 2b\,\mathcal{H}^{n-1}(S(u)) & \text{if } u \in GSBV(\Omega), \\ +\infty & \text{otherwise.} \end{cases} \tag{5.3}$$

Our main result is the following theorem, which will follow from Propositions 5.12 and 5.13.

**Theorem 5.1** *The family $(F_\varepsilon)$ $\Gamma$-converges to $F$ in $L^1(\Omega)$ as $\varepsilon \to 0^+$.*

### 5.1.1 *Estimate from below of the volume term*

In order to estimate from below the $\Gamma$-limit it is not possible to proceed using the slicing procedure. We will nevertheless follow in a way the steps outlined in Section 4.1.1, first localizing the functionals, then obtaining estimates from below by a family of integral functionals, and eventually recovering the best lower estimate by a supremum of measures.

For every open set $A \subseteq \Omega$ and for every $\varepsilon > 0$ we consider the functional

$$F_\varepsilon(u,A)=\begin{cases}\dfrac{1}{\varepsilon}\displaystyle\int_A f\Big(\varepsilon \fint_{B_\varepsilon(x)\cap\Omega}|\nabla u(y)|^2\,dy\Big)\,dx & \text{if } u\in H^1(\Omega)\\ +\infty & \text{if } u\in L^1(\Omega)\setminus H^1(\Omega),\end{cases} \tag{5.4}$$

which provides a "localization" of the functional $F_\varepsilon(u)$.

The aim of this section is to prove an estimate from below of $F_\varepsilon(u,A)$ in terms of $\int_A |\nabla v|^2\,dx$, where $v$ is a function in $SBV(A)$ which is very close to $u$ in $L^1(A)$. This fact will be crucial to obtain a good lower bound for the $\Gamma$-limit of $(F_\varepsilon)$. We use the notation

$$A_\rho=\{x\in A:\operatorname{dist}(x,\partial A)>\rho\} \tag{5.5}$$

for every open set $A\subseteq\Omega$ and for every $\rho>0$. Moreover, if $\eta>0$ and $\alpha\in\mathbf{Z}^n$

$$Q_\eta^\alpha=\eta\alpha+(-\eta/2,\eta/2)^n$$

and $Q_\eta=Q_\eta^0=(-\eta/2,\eta/2)^n$.

**Proposition 5.2** *Let $f=\min\{at,b\}$. Let $A$ be an open subset of $\Omega$, and let $u\in H^1(\Omega)\cap L^\infty(\Omega)$. For every $\varepsilon>0$ and $\delta>0$ there exists a function $v\in SBV(A)\cap L^\infty(A)$ such that*

$$\begin{aligned}(1-\delta)a\int_A|\nabla v(x)|^2\,dx&\le F_\varepsilon(u,A),\\ \mathcal{H}^{n-1}(S(v)\cap A_{6\varepsilon})&\le c\,F_\varepsilon(u,A),\\ \|v\|_{L^\infty(A)}&\le\|u\|_{L^\infty(A)},\\ \|v-u\|_{L^1(A_{6\varepsilon})}&\le c\varepsilon\,F_\varepsilon(u,A)\,\|u\|_{L^\infty(A)},\end{aligned}$$

*where $c$ is a constant depending on $n$, $\delta$ and $f$ only.*

Note that the coefficient of $\int_A|\nabla v|^2\,dx$ can be chosen arbitrarily close to 1. To prove Proposition 5.2 we need the following two lemmas.

**Lemma 5.3** *Let $\psi:\mathbf{R}^n\to\mathbf{R}$ be a measurable function with compact support. For every $\eta>0$ there exists $x_0\in(-\eta/2,\eta/2)^n$ such that*

$$\int_{\mathbf{R}^n}\psi(x)\,dx\ge\sum_{\alpha\in\mathbf{Z}^n}\eta^n\,\psi(x_0+\eta\alpha).$$

**Proof** We can write

$$\int_{\mathbf{R}^n}\psi(x)\,dx=\sum_{\alpha\in\mathbf{Z}^n}\int_{Q_\eta}\psi(x+\eta\alpha)\,dx=\int_{Q_\eta}\Psi(x)\,dx$$

where $\Psi:Q_\eta\to\mathbf{R}$ is the function defined by $\Psi(x)=\sum_{\alpha\in\mathbf{Z}^n}\psi(x+\eta\alpha)$ (note that the sum runs over a finite number of indices). The conclusion follows applying the Mean Value Theorem to the function $\Psi$. □

**Lemma 5.4** *For every $\delta > 0$ there exists $s \in (0, 1/\sqrt{n})$, only depending on $n$ and $\delta$, such that*

$$\sum_{\alpha \in \mathbf{Z}^n} \frac{\eta^n}{|B_\varepsilon|} \chi_{B_\varepsilon(\eta\alpha)}(x) \geq 1 - \delta \tag{5.6}$$

*for every $x \in \mathbf{R}^n$ and for every $\varepsilon > 0$ and $\eta > 0$ with $\eta \leq s\varepsilon$.*

**Proof** We may assume that $0 < \delta < 1$. For every $x \in \mathbf{R}^n$ and for every $\eta > 0$ let $A_\eta(x)$ be the union of all cubes $Q_\eta^\alpha$, $\alpha \in \mathbf{Z}^n$, which are contained in $B_1(x)$, and let $r_\eta = 1 - \eta\sqrt{n}$. Since $A_\eta(x) \supseteq B_{r_\eta}(x)$, for $0 < \eta \leq (1 - (1-\delta)^{1/n})/\sqrt{n} =: s$ we have

$$\sum_{\alpha \in \mathbf{Z}^n} \frac{\eta^n}{|B_1|} \chi_{B_1(\eta\alpha)}(x) = \sum_{\alpha \in \mathbf{Z}^n} \frac{\eta^n}{|B_1|} \chi_{B_1(x)}(\eta\alpha) \tag{5.7}$$

$$\geq \frac{\eta^n}{|B_1|} \# \Big\{ \alpha \in \mathbf{Z}^n : \eta\alpha + (-\eta/2, \eta/2)^n \subseteq B_1(x) \Big\}$$

$$= \frac{|A_\eta(x)|}{|B_1|} \geq \frac{|B_{r_\eta}|}{|B_1|} = (1 - \eta\sqrt{n})^n \geq 1 - \delta\,.$$

This shows that (5.6) holds for $\varepsilon = 1$ and $\eta \leq s$. Moreover, a simple scaling argument shows that for every $\varepsilon > 0$ and $\eta > 0$

$$\sum_{\alpha \in \mathbf{Z}^n} \frac{\eta^n}{|B_\varepsilon|} \chi_{B_\varepsilon(\eta\alpha)}(x) = \sum_{\alpha \in \mathbf{Z}^n} \frac{\eta^n}{\varepsilon^n |B_1|} \chi_{B_\varepsilon(\eta\alpha)}(x) = \sum_{\alpha \in \mathbf{Z}^n} \frac{(\eta/\varepsilon)^n}{|B_1|} \chi_{B_1((\eta/\varepsilon)\alpha)}(x/\varepsilon)\,.$$

If we apply (5.7) with $\eta$ and $x$ replaced by $\eta/\varepsilon$ and $x/\varepsilon$ we obtain (5.6) in the general case. □

**Proof of Proposition 5.2** It is not restrictive to deal with the case $a = 1$ only. Fix $\varepsilon > 0$ and $\delta > 0$. Let $\psi_\varepsilon : \mathbf{R}^n \to \mathbf{R}$ be defined by

$$\psi_\varepsilon(x) = f\Big(\varepsilon \fint_{B_\varepsilon(x)\cap\Omega} |\nabla u(y)|^2\, dy\Big)$$

for $x \in A$, and by $\psi_\varepsilon(x) = 0$ for $x \in \mathbf{R}^n \setminus A$, so that

$$F_\varepsilon(u, A) = \frac{1}{\varepsilon} \int_{\mathbf{R}^n} \psi_\varepsilon(x)\, dx\,. \tag{5.8}$$

Applying Lemma 5.4 we fix $s \in (0, 1/\sqrt{n})$, depending on $n$ and $\delta$ only, such that for every $x \in \mathbf{R}^n$ we have

$$\sum_{\alpha \in \mathbf{Z}^n} \frac{\eta^n}{|B_\varepsilon|} \chi_{B_\varepsilon(\eta\alpha)}(x) \geq (1 - \delta)\,, \tag{5.9}$$

where $\eta = s\,\varepsilon$. By Lemma 5.3 we have

$$\int_{\mathbf{R}^n} \psi_\varepsilon(x)\,dx \geq \sum_{\alpha\in\mathbf{Z}^n} \eta^n \psi_\varepsilon(x_0+\eta\alpha) \tag{5.10}$$

for a suitable $x_0 \in \mathbf{R}^n$. Up to a change of variables we may assume that $x_0 = 0$.

For every $\rho > 0$ let

$$I_\rho = \{\alpha \in \mathbf{Z}^n : \eta\alpha \in A_\rho\}\,. \tag{5.11}$$

For every $\alpha \in \mathbf{Z}^n$ and for every $I \subseteq \mathbf{Z}^n$ we set

$$B^\alpha_\rho = B_\rho(\eta\alpha)\,, \qquad Q_\eta(I) = \mathrm{int}\Big(\bigcup_{\beta\in I} \overline{Q}^\beta_\eta\Big)\,, \tag{5.12}$$

where $\mathrm{int}(E)$ and $\overline{E}$ denote the interior and the closure of the set $E$, respectively. Since $\varepsilon > \eta\sqrt{n}$ we have

$$Q^\alpha_\eta \subseteq B^\alpha_\varepsilon\,, \qquad \bigcup_{\alpha\in\mathbf{Z}^n} B^\alpha_\varepsilon = \mathbf{R}^n\,, \qquad A_{6\varepsilon} \subseteq Q_\eta(I_{5\varepsilon})\,. \tag{5.13}$$

From (5.8) and (5.10) we obtain

$$F_\varepsilon(u,A) \geq \frac{1}{\varepsilon}\sum_{\alpha\in I_\varepsilon} \eta^n \psi_\varepsilon(\eta\alpha) = \sum_{\alpha\in I_\varepsilon} \frac{\eta^n}{\varepsilon} f\Big(\varepsilon \fint_{B^\alpha_\varepsilon} |\nabla u(y)|^2\,dy\Big)\,. \tag{5.14}$$

For every $\rho \geq r > 0$ and for every $t > 0$ we define the sets of indices

$$\begin{aligned} I^r_\rho(t) &= \{\alpha \in I_\rho : r\fint_{B^\alpha_r} |\nabla u(y)|^2\,dy < t\}\,, \\ J^r_\rho(t) &= \{\alpha \in I_\rho : r\fint_{B^\alpha_r} |\nabla u(y)|^2\,dy \geq t\}\,. \end{aligned} \tag{5.15}$$

Let $C = 3^{-n}b$. Note that, if $\alpha \in I^{3\varepsilon}_{3\varepsilon}(C)$, $\beta \in \mathbf{Z}^n$, and $Q^\alpha_\eta \cap B^\beta_\varepsilon \neq \emptyset$, then $B^\beta_\varepsilon \subseteq B^\alpha_{3\varepsilon}$ and $\beta \in I^\varepsilon_\varepsilon(b)$. In particular, if $\alpha \in I^{3\varepsilon}_{3\varepsilon}(C)$, from (5.9) we obtain

$$\sum_{\beta\in I^\varepsilon_\varepsilon(b)} \frac{\eta^n}{|B_\varepsilon|}\chi_{B^\beta_\varepsilon}(x) \geq (1-\delta)\chi_{Q^\alpha_\eta}(x)\,. \tag{5.16}$$

Let $v : A \to \mathbf{R}$ be the function defined by

$$v(x) = \begin{cases} u(x) & \text{if } x \in Q_\eta(I^{3\varepsilon}_{5\varepsilon}(C)) \\ 0 & \text{otherwise.} \end{cases}$$

Since $u \in H^1(\Omega) \cap L^\infty(\Omega)$, clearly $v \in SBV(A) \cap L^\infty(A)$ and $\|v\|_{L^\infty(A)} \leq \|u\|_{L^\infty(A)}$. From (5.14) and (5.16) we obtain

$$F_\varepsilon(u,A) \geq \sum_{\alpha\in I^\varepsilon_\varepsilon(b)} \frac{\eta^n}{\varepsilon} f\Big(\varepsilon \fint_{B^\alpha_\varepsilon} |\nabla u(x)|^2\,dx\Big)$$

$$= \sum_{\alpha \in I_\varepsilon^\varepsilon(b)} \frac{\eta^n}{|B_\varepsilon|} \int_{B_\varepsilon^\alpha} |\nabla u(x)|^2 \, dx \geq$$
$$\geq (1-\delta) \sum_{\alpha \in I_{5\varepsilon}^{3\varepsilon}(C)} \int_{Q_\eta^\alpha} |\nabla v(x)|^2 \, dx = (1-\delta) \int_A |\nabla v(x)|^2 \, dx \, .$$

As $A_{6\varepsilon} \subseteq Q_\eta(I_{5\varepsilon})$ by (5.13), in order to estimate $\|v-u\|_{L^1(A_{6\varepsilon})}$ it suffices to evaluate the size of $Q_\eta(J_{5\varepsilon}^{3\varepsilon}(C))$. If $\beta \in J_{5\varepsilon}^{3\varepsilon}(C)$, then

$$3\varepsilon \fint_{B_{3\varepsilon}^\beta} |\nabla u(x)|^2 \, dx \geq C \, ;$$

by subadditivity and by (5.13) this implies that

$$\frac{\varepsilon}{3^{n-1}} \sum_{B_\varepsilon^\alpha \cap B_{3\varepsilon}^\beta \neq \emptyset} \fint_{B_\varepsilon^\alpha} |\nabla u(x)|^2 \, dx \geq C \, ,$$

so that there exists $\alpha \in \mathbf{Z}^n$ such that $B_\varepsilon^\alpha \cap B_{3\varepsilon}^\beta \neq \emptyset$ and

$$\varepsilon \fint_{B_\varepsilon^\alpha} |\nabla u(x)|^2 \, dx \geq k \, , \tag{5.17}$$

where $k = 3^{n-1} C / \gamma_n^s = b / 3\gamma_n^s$ and

$$\gamma_n^s = \#\{\alpha \in \mathbf{Z}^n : B_{1/s}(\alpha) \cap B_{3/s}(0) \neq \emptyset\} = \#\{\alpha \in \mathbf{Z}^n : B_\varepsilon^\alpha \cap B_{3\varepsilon}^\beta \neq \emptyset\} \, .$$

Since $\beta \in I_{5\varepsilon}$ and $B_\varepsilon^\alpha \cap B_{3\varepsilon}^\beta \neq \emptyset$, we have that $\alpha \in I_\varepsilon$ and (5.17) gives $\alpha \in J_\varepsilon^\varepsilon(k)$. This implies that for every $\beta \in J_{5\varepsilon}^{3\varepsilon}(C)$ there exists $\alpha \in J_\varepsilon^\varepsilon(k)$ such that $B_\varepsilon^\alpha \cap B_{3\varepsilon}^\beta \neq \emptyset$. Since for every $\alpha \in J_\varepsilon^\varepsilon(k)$ we have

$$\gamma_n^s = \#\{\beta \in \mathbf{Z}^n : B_\varepsilon^\alpha \cap B_{3\varepsilon}^\beta \neq \emptyset\} \, ,$$

we deduce that

$$\# J_{5\varepsilon}^{3\varepsilon}(C) \leq \gamma_n^s \, \# J_\varepsilon^\varepsilon(k) \, . \tag{5.18}$$

Let $c = \gamma_n^s / f(k)$. From (5.18) and (5.14) it follows that

$$\frac{1}{\varepsilon} \# J_{5\varepsilon}^{3\varepsilon}(C) \, \eta^n \leq \frac{\gamma_n^s \, f(k)}{\varepsilon \, f(k)} \# J_\varepsilon^\varepsilon(k) \, \eta^n$$
$$\leq c \frac{1}{\varepsilon} \sum_{\alpha \in I_\varepsilon} \eta^n f\Big(\varepsilon \fint_{B_\varepsilon^\alpha} |\nabla u(x)|^2 \, dx\Big) \leq c \, F_\varepsilon(u, A) \, . \tag{5.19}$$

Since $A_{6\varepsilon} \subseteq Q_\eta(I_{5\varepsilon})$ by (5.13), we obtain

$$\int_{A_{6\varepsilon}} |v-u| \, dx \leq \eta^n \# J_{5\varepsilon}^{3\varepsilon}(C) \, \|u\|_{L^\infty(A)} \leq c \, \varepsilon \, F_\varepsilon(u, A) \, \|u\|_{L^\infty(A)} \, .$$

Eventually we estimate $\mathcal{H}^{n-1}(S(v)\cap A_{6\varepsilon})$. By the definition of $v$ we have

$$S(v)\cap A_{6\varepsilon}\subseteq \bigcup_{\alpha\in J^{3\varepsilon}_{5\varepsilon}(C)} \partial Q^{\alpha}_{\eta}\,,$$

so that by (5.19)

$$\mathcal{H}^{n-1}(S(v)\cap A_{6\varepsilon})\le \# J^{3\varepsilon}_{5\varepsilon}(C)\,2n\,\eta^{n-1}\le 2n\,\frac{\varepsilon}{\eta}\,c\,F_\varepsilon(u,A)=\frac{2n}{s}\,c\,F_\varepsilon(u,A)\,,$$

and the estimate is obtained. □

5.1.2 *Estimate from below of the surface term*

In this section, given $\xi\in S^{n-1}$ and $u\in H^1(\Omega)\cap L^\infty(\Omega)$, we prove an estimate from below of $F_\varepsilon(u,A)$ in terms of $\int_{S(w)\cap A_{8\sigma}}|\langle\nu_w,\xi\rangle|\,d\mathcal{H}^{n-1}$, where $w$ is a function in $SBV(A)$ which coincides with $u$ in a large subset of $A$, and $A_{8\sigma}$ is defined by (5.5) with $\sigma=\varepsilon\sqrt{n}$. Note that the coefficient of the surface integral appearing in the estimate can be chosen arbitrarily close to $2b$. This fact will be crucial to obtain a good lower bound for the $\Gamma$-limit of $(F_\varepsilon)$.

**Proposition 5.5** *Let $f(t)=\min\{at,b\}$. Let $A$ be an open subset of $\Omega$, and let $u\in H^1(\Omega)\cap L^\infty(\Omega)$. For every $\varepsilon>0$, $\delta>0$, $\xi\in S^{n-1}$ there exist a function $w\in SBV(A)\cap L^\infty(A)$ and a Borel set $E\subseteq A$ such that*

$$2\,(1-\delta)\,b\int_{S(w)\cap A_{8\sigma}}|\langle\nu_w,\xi\rangle|\,d\mathcal{H}^{n-1}\le F_\varepsilon(u,A)\,,$$

$$\int_A|\nabla w|^2\,dx\le c\,F_\varepsilon(u,A)\,,\qquad \mathcal{H}^{n-1}(S(w)\cap A_{8\sigma})\le c\,F_\varepsilon(u,A)\,,$$

$$\|w-u\|_{L^1(E)}\le c\,\varepsilon\,F_\varepsilon(u,A)\,\|u\|_{L^\infty(A)}\,,\qquad \|w\|_{L^\infty(A)}\le\|u\|_{L^\infty(A)}\,,$$

$$|E\cap B_{2\sigma}(x)|\ge|B_{\varepsilon/2}|\ \text{for all}\ x\in A_{2\sigma}\quad,$$

*where $\sigma=\varepsilon\sqrt{n}$ and $c$ is a constant depending on $n$, $p$, $\delta$ and $b$ only.*

To prove Proposition 5.5 we need an extension result, with a careful estimate of the extension constant. For every $I\subseteq\mathbf{Z}^n$ and for every integer $k>0$ we define

$$\begin{aligned} e_kI&=\{\alpha\in\mathbf{Z}^n:\min_{\beta\in I}|\alpha-\beta|_\infty\le k\}\,,\\ i_kI&=\{\alpha\in I:\min_{\beta\in\mathbf{Z}^n\setminus I}|\alpha-\beta|_\infty>k\}\,,\\ b_kI&=\{\alpha\in I:\min_{\beta\in\mathbf{Z}^n\setminus I}|\alpha-\beta|_\infty\le k\}\,,\end{aligned}\tag{5.20}$$

where $|\alpha|_\infty=\max\{|\alpha^1|,\dots,|\alpha^n|\}$ for every $\alpha=(\alpha^1,\dots,\alpha^n)\in\mathbf{Z}^n$. Given $\eta>0$ and $I\subseteq\mathbf{Z}^n$, let $Q_\eta(I)$ be the set introduced in (5.12). Moreover, we define

$$B^\eta_\rho(I)=\bigcup_{\alpha\in I}B_\rho(\eta\alpha)\tag{5.21}$$

for every $\rho > 0$. We are now in a position to prove the following lemma.

**Lemma 5.6** *Let $1/2 < s < \sqrt{2}/2$. There exists a constant $C$, depending on $n$ and $s$ only, with the following property: for every $\eta > 0$, $I \subseteq \mathbf{Z}^n$, $u \in H^1(B^\eta_{s\eta}(I))$ there exists $v \in H^1(Q_\eta(I))$ such that $v = u$ in $Q_\eta(I) \cap B^\eta_{s\eta}(I)$ and*

$$\int_{Q_\eta(J)} |\nabla v|^2\,dx \leq C \int_{B^\eta_{s\eta}(I\cap e_2 J)} |\nabla u|^2\,dx \tag{5.22}$$

*for every $J \subseteq i_1 I$. Moreover, if in addition $u \in L^\infty(B^\eta_{s\eta}(I))$, then we also have $\|v\|_{L^\infty(Q_\eta(I))} \leq \|u\|_{L^\infty(B^\eta_{s\eta}(I))}$.*

**Proof** Throughout the proof the letter $c$ stands for a positive constant depending at most on $n$ and $s$, whose value may change from formula to formula. Let $\eta > 0$, let $I \subseteq \mathbf{Z}^n$, and let $u \in H^1(B^\eta_\varepsilon(I))$, where $\varepsilon = s\eta$. For every $\alpha \in \mathbf{Z}^n$ let $B^\alpha_\varepsilon$ and $Q^\alpha_\eta$ be the sets defined in (5.12) and let $\widehat{Q}^\alpha_\eta = \eta\alpha + (-\eta,\eta)^n$ and $\widetilde{Q}^\alpha_\eta = \eta\alpha + (-3\eta/2, 3\eta/2)^n$. As $1/2 < s < \sqrt{2}/2$, we have $B^\alpha_\varepsilon \cap B^\beta_\varepsilon \neq \emptyset$ if and only if $|\alpha-\beta|_1 \leq 1$, where $|\gamma|_1 = \sum_i |\gamma^i|$ for every $\gamma = (\gamma^1,\ldots,\gamma^n) \in \mathbf{Z}^n$. This implies that the sets $\widehat{Q}^\alpha_\eta \cap B^\eta_\varepsilon(I)$ have a Lipschitz boundary. Therefore for every $\alpha \in \mathbf{Z}^n$ there exists an extension $v^\alpha_\eta \in H^1(\widetilde{Q}^\alpha_\eta)$ of $u|_{\widehat{Q}^\alpha_\eta \cap B^\eta_\varepsilon(I)}$ such that

$$\int_{\widetilde{Q}^\alpha_\eta} |\nabla v^\alpha_\eta|^2\,dx \leq c \int_{\widehat{Q}^\alpha_\eta \cap B^\eta_\varepsilon(I)} |\nabla u|^2\,dx\,. \tag{5.23}$$

Note that, by the geometry of the problem, the number of equivalence classes of the sets $\widehat{Q}^\alpha_\eta \cap B^\eta_\varepsilon(I)$ modulo similarities is bounded independently of $I$, $\alpha$, and $\eta$, so that $c$ can be chosen indeed depending on $n$ and $s$ only.

We now define a suitable partition of unity. Fix $\varphi^0_\eta \in C^\infty_c(\widehat{Q}^0_\eta)$ with $\varphi^0_\eta = 1$ in $Q^0_\eta$, $0 \leq \varphi^0_\eta \leq 1$ in $\widehat{Q}^0_\eta$, and $|\nabla\varphi^0_\eta| \leq c/\eta$ in $\widehat{Q}^0_\eta$. We set $\varphi^\alpha_\eta(x) = \varphi^0_\eta(x-\eta\alpha)$ for every $\alpha \in \mathbf{Z}^n$, so that $\varphi^\alpha_\eta \in C^\infty_c(\widehat{Q}^\alpha_\eta)$, $\varphi^\alpha_\eta = 1$ in $Q^\alpha_\eta$, $0 \leq \varphi^\alpha_\eta \leq 1$ in $\widehat{Q}^\alpha_\eta$, and $|\nabla\varphi^\alpha_\eta| \leq c/\eta$ in $\widehat{Q}^\alpha_\eta$. Finally we define $\psi^\alpha_\eta \in C^\infty(Q_\eta(I))$ by $\psi^\alpha_\eta = \varphi^\alpha_\eta/\varphi_\eta$, where $\varphi_\eta = \sum_{\beta\in I}\varphi^\beta_\eta$. Then we have $\sum_{\alpha\in I}\psi^\alpha_\eta = 1$ and $|\nabla\psi^\alpha_\eta| \leq c/\eta$ in $Q_\eta(I)$.

We define $v \in H^1(Q_\eta(I))$ by

$$v = \sum_{\beta\in I} \psi^\beta_\eta v^\beta_\eta\,.$$

Since $v^\beta_\eta = u$ on $\widehat{Q}^\beta_\eta \cap B^\eta_\varepsilon(I)$ we have $\psi_\beta v^\beta_\eta = \psi_\beta u$ in $Q_\eta(I) \cap B^\eta_\varepsilon(I)$; hence, $v = u$ in $Q_\eta(I) \cap B^\eta_\varepsilon(I)$.

Let $J \subseteq i_1 I$, let $\alpha \in J$, and let $I^\alpha_1 = \{\beta \in I : |\beta-\alpha|_\infty \leq 1\}$. Note that in $Q^\alpha_\eta$ the sum which appears in the definition of $v$ can be restricted to $I^\alpha_1$. Moreover, since $\sum_{\beta\in I^\alpha_1}\psi_\beta = 1$ in $Q^\alpha_\eta$, we also have $\sum_{\beta\in I^\alpha_1}\nabla\psi^\beta_\eta = 0$ in $Q^\alpha_\eta$. Hence

$$\nabla v = \sum_{\beta\in I^\alpha_1} \psi_\beta \nabla v^\beta_\eta + \sum_{\beta\in I^\alpha_1} (v^\beta_\eta - v^\alpha_\eta)\nabla\psi_\beta$$

in $Q_\eta^\alpha$, so that by (5.23)

$$\int_{Q_\eta^\alpha} |\nabla v|^2\,dx \le c \sum_{\beta\in I_1^\alpha} \int_{\widetilde{Q}_\eta^\beta} |\nabla v_\eta^\beta|^2\,dx + c\sum_{\beta\in I_1^\alpha} \int_{Q_\eta^\alpha} |v_\eta^\beta - v_\eta^\alpha|^2 |\nabla \psi_\eta^\beta|^2\,dx$$

$$\le c \sum_{\beta\in I_1^\alpha} \int_{\widehat{Q}_\eta^\beta \cap B_\varepsilon^\eta(I)} |\nabla u|^2\,dx + \frac{c}{\eta^2} \sum_{\beta\in I_1^\alpha} \int_{Q_\eta^\alpha} |v_\eta^\beta - v_\eta^\alpha|^2\,dx\,. \qquad (5.24)$$

If $\lambda,\,\mu \in I_1^\alpha$ and $|\lambda-\mu|_1 \le 1$, then $Q_\eta^\alpha \subseteq \widetilde{Q}_\eta^\lambda \cap \widetilde{Q}_\eta^\mu$ and $B_\varepsilon^\lambda \cap B_\varepsilon^\mu \neq \emptyset$. Since $v_\eta^\lambda = u = v_\eta^\mu$ in $B_\varepsilon^\lambda \cap B_\varepsilon^\mu$, by Poincaré's inequality we have

$$\int_{Q_\eta^\alpha} |v_\eta^\lambda - v_\eta^\mu|^2\,dx \le \int_{\widetilde{Q}_\eta^\lambda \cap \widetilde{Q}_\eta^\mu} |v_\eta^\lambda - v_\eta^\mu|^2\,dx \le c\,\eta^2 \int_{\widetilde{Q}_\eta^\lambda \cap \widetilde{Q}_\eta^\mu} |\nabla v_\eta^\lambda - \nabla v_\eta^\mu|^2\,dx\,. \quad (5.25)$$

Since $\alpha \in i_1 I$, for every $\beta \in I_1^\alpha$ there exists a family $\gamma_0, \gamma_1, \ldots, \gamma_n$ of $n+1$ elements of $I_1^\alpha$ such that

$$\gamma_0 = \alpha\,, \qquad \gamma_n = \beta\,, \qquad |\gamma_i - \gamma_{i-1}|_1 \le 1 \quad \text{for } i = 1,\ldots,n\,.$$

From (5.25) and (5.23) it follows that for every $\beta \in I_1^\alpha$

$$\int_{Q_\eta^\alpha} |v_\eta^\beta - v_\eta^\alpha|^2\,dx \le c \sum_{i=1}^n \int_{Q_\eta^\alpha} |v_\eta^{\gamma_i} - v_\eta^{\gamma_{i-1}}|^2\,dx \le c \sum_{\gamma\in I_1^\alpha} \int_{\widehat{Q}_\eta^\gamma \cap B_\varepsilon^\eta(I)} |\nabla u|^2\,dx\,,$$

so that by (5.24) we have for every $\alpha \in J$

$$\int_{Q_\eta^\alpha} |\nabla v|^2\,dx \le c \int_{B_\varepsilon^\eta(I_2^\alpha)} |\nabla u|^2\,dx\,,$$

where $I_2^\alpha = \{\beta \in I : |\beta - \alpha|_\infty \le 2\}$. By taking the sum for $\alpha \in J$ we obtain (5.22).

If $u \in L^\infty(B_\varepsilon^\eta(I))$, first we construct $v$ as above, and then we take $(v \wedge k) \vee (-k)$, with $k = \|u\|_{L^\infty(B_\varepsilon^\eta(I))}$. □

To prove Proposition 5.5 we need a generalization of Lemma 5.6, which allows to extend (5.22) to sets $J \subseteq I$ which are not contained in $i_1 I$. For every cube $Q_\eta^\alpha$ we define $\partial_1 Q_\eta^\alpha$ as the union of its $n-1$ dimensional faces which are orthogonal to the hyperplane $\{x^1 = 0\}$.

**Lemma 5.7** *Let $1/2 < s < \sqrt{2}/2$. There exists a constant $C$, depending only on $n$ and $s$, with the following property: for every $\eta > 0$, $I \subseteq \mathbf{Z}^n$, $u \in H^1(B_{s\eta}^\eta(I))$ there exists $w \in SBV(Q_\eta(I))$ such that $w = u$ in $Q_\eta(I) \cap B_{s\eta}^\eta(I)$,*

$$S(w) \subseteq \bigcup_{\alpha\in b_3 I} \partial_1 Q_\eta^\alpha \qquad \textit{up to an } \mathcal{H}^{n-1}\textit{-negligible set,} \qquad (5.26)$$

*and*

$$\int_{Q_\eta(I)} |\nabla w|^2\,dx \le C \int_{B^\eta_{s\eta}(I)} |\nabla u|^2\,dx\,. \tag{5.27}$$

*Moreover, if* $u \in L^\infty(B^\eta_{s\eta}(I))$, *then we also have* $\|w\|_{L^\infty(Q_\eta(I))} \le \|u\|_{L^\infty(B^\eta_{s\eta}(I))}$.

**Proof** Throughout the proof the letter $c$ stands for a positive constant depending at most on $n$ and $s$, whose value may change from formula to formula. Let $\eta > 0$, let $I \subseteq \mathbf{Z}^n$, and let $u \in H^1(B^\eta_\varepsilon(I))$, where $\varepsilon = s\eta$. We set

$$\widehat{R}^0_\eta = (-\eta,\eta) \times (-\eta/2,\eta/2)^{n-1}\,, \quad \widetilde{R}^0_\eta = (-3\eta/2, 3\eta/2) \times (-\eta/2,\eta/2)^{n-1}\,,$$
$$T^0_\eta = \mathbf{R} \times (-\eta/2,\eta/2)^{n-1}\,,$$

and for every $\alpha \in \mathbf{Z}^n$ we define $\widehat{R}^\alpha_\eta = \eta\alpha + \widehat{R}^0_\eta$, $\widetilde{R}^\alpha_\eta = \eta\alpha + \widetilde{R}^0_\eta$, and $T^\alpha_\eta = \eta\alpha + T^0_\eta$. By Lemma 5.6 there exists $v \in H^1(Q_\eta(I))$ such that $v = u$ in $Q_\eta(I) \cap B^\eta_\varepsilon(I)$ and

$$\int_{\widetilde{R}^\alpha_\eta} |\nabla v|^2\,dx \le c \int_{B^\eta_\varepsilon(I^\alpha_3)} |\nabla u|^2\,dx \tag{5.28}$$

for every $\alpha \in i_2 I$, where $I^\alpha_3 = \{\beta \in I : |\beta - \alpha|_\infty \le 3\}$.

As $1/2 < s < \sqrt{2}/2$, the set $\widehat{R}^\alpha_\eta \cap B^\eta_\varepsilon(I)$ has a Lipschitz boundary. Therefore for every $\alpha \in b_2 I$ there exists an extension $w^\alpha_\eta \in H^1(\widetilde{R}^\alpha_\eta)$ of $u|_{\widehat{R}^\alpha_\eta \cap B^\eta_\varepsilon(I)}$ such that

$$\int_{\widetilde{R}^\alpha_\eta} |\nabla w^\alpha_\eta|^2\,dx \le c \int_{\widehat{R}^\alpha_\eta \cap B^\eta_\varepsilon(I)} |\nabla u|^2\,dx\,. \tag{5.29}$$

Note that, by the geometry of the problem, the number of equivalence classes of the sets $\widehat{R}^\alpha_\eta \cap B^\eta_\varepsilon(I)$ modulo similarities is bounded independently of $I$, $\alpha$, and $\eta$, so that $c$ can be chosen indeed depending only on $n$ and $s$. For every $\alpha \in i_2 I$ we set $w^\alpha_\eta = v$ in $\widetilde{R}^\alpha_\eta$, so that $w^\alpha_\eta \in H^1(\widetilde{R}^\alpha_\eta)$ for every $\alpha \in I$ and, by (5.28) and (5.29),

$$\int_{\widetilde{R}^\alpha_\eta} |\nabla w^\alpha_\eta|^2\,dx \le c \int_{\widehat{R}^\alpha_\eta \cap B^\eta_\varepsilon(I)} |\nabla u|^2\,dx \le c \int_{B^\eta_\varepsilon(I^\alpha_3)} |\nabla u|^2\,dx \tag{5.30}$$

for every $\alpha \in I$.

We now define a suitable partition of unity composed of functions which may be discontinuous on the sets $\partial_1 Q^\alpha_\eta$. For every $\alpha \in \mathbf{Z}^n$ let $\varphi^\alpha_\eta$ be the function defined in the proof of Lemma 5.6 and let $\xi^\alpha_\eta = \varphi^\alpha_\eta$ in the closure of the rectangle $\widetilde{R}^\alpha_\eta$, while $\xi^\alpha_\eta = 0$ elsewhere in $\mathbf{R}^n$. Then $\xi^\alpha_\eta \in C^\infty(T^\beta_\eta)$ for every $\beta \in \mathbf{Z}^n$, $\xi^\alpha_\eta = 1$ in $\overline{Q}^\alpha_\eta$, $0 \le \xi^\alpha_\eta \le 1$ in $\widehat{R}^\alpha_\eta$, $\xi^\alpha_\eta = 0$ a.e. in $\mathbf{R}^n \setminus \widehat{R}^\alpha_\eta$, and $|\nabla \xi^\alpha_\eta| \le c/\eta$ a.e. in $\mathbf{R}^n$. Finally we define $\zeta^\alpha_\eta : Q_\eta(I) \to \mathbf{R}$ by $\zeta^\alpha_\eta = \xi^\alpha_\eta / \xi_\eta$, where $\xi_\eta = \sum_{\beta \in I} \xi^\beta_\eta$. Then $\zeta^\alpha_\eta \in C^\infty(T^\beta_\eta \cap Q_\eta(I))$ for every $\beta \in I$. Moreover $\sum_{\alpha \in I} \zeta^\alpha_\eta = 1$ in $Q_\eta(I)$ and $|\nabla \zeta^\alpha_\eta| \le c/\eta$ a.e. in $Q_\eta(I)$.

We define $w : Q_\eta(I) \to \mathbf{R}$ by $w = \sum_{\beta\in I} \zeta_\eta^\beta w_\eta^\beta$. Then $w \in H^1(T_\eta^\beta \cap Q_\eta(I))$ for every $\beta \in I$; hence, $w \in SBV(Q_\eta(I))$. As $w_\eta^\beta = u$ a.e. in $\widehat{R}_\eta^\beta \cap B_\varepsilon^\eta(I)$, we have $\zeta_\beta w_\eta^\beta = \zeta_\beta u$ a.e. in $Q_\eta(I) \cap B_\varepsilon^\eta(I)$; hence, $w = u$ a.e. in $Q_\eta(I) \cap B_\varepsilon^\eta(I)$.

For every $\alpha \in I$ let $J_1^\alpha = \{\alpha - e_1, \alpha, \alpha + e_1\} \cap I$, where $e_1 = (1, 0, \ldots, 0)$. Note that in $Q_\eta^\alpha$ the sum which appears in the definition of $w$ can be restricted to $J_1^\alpha$. If $\alpha \in i_3 I$, then $J_1^\alpha \subseteq i_2 I$, so that $w_\eta^\beta = v$ in $Q_\eta^\alpha$ for every $\beta \in J_1^\alpha$. Since $\sum_{\beta\in J_1^\alpha} \zeta_\beta = 1$ in $Q_\eta^\alpha$, we have $w = v$ in $Q_\eta^\alpha$ for every $\alpha \in i_3 I$; hence, $w = u$ a.e. in $Q_\eta(i_3 I)$. This shows that $u \in H^1(Q_\eta(i_3 I))$ and, consequently, $S(w) \subseteq b_3 I$ up to an $\mathcal{H}^{n-1}$-negligible set. As $u \in H^1(T_\eta^\beta \cap Q_\eta(I))$ for every $\beta \in I$, we have proved (5.26).

As $\sum_{\beta\in J_1^\alpha} \zeta_\beta = 1$ in $Q_\eta^\alpha$, we also have $\sum_{\beta\in J_1^\alpha} \nabla\zeta_\eta^\beta = 0$ a.e. in $Q_\eta^\alpha$. Hence

$$\nabla w = \sum_{\beta\in J_1^\alpha} \zeta_\beta \nabla w_\eta^\beta + \sum_{\beta\in J_1^\alpha} (w_\eta^\beta - w_\eta^\alpha)\nabla\zeta_\beta$$

in $Q_\eta^\alpha$, so that by (5.30)

$$\begin{aligned}\int_{Q_\eta^\alpha} |\nabla w|^2\,dx &\le c\sum_{\beta\in J_1^\alpha} \int_{\widetilde{R}_\eta^\beta} |\nabla w_\eta^\beta|^2\,dx + c\sum_{\beta\in J_1^\alpha} \int_{Q_\eta^\alpha} |w_\eta^\beta - w_\eta^\alpha|^2 |\nabla\zeta_\eta^\beta|^2\,dx \\ &\le c\sum_{\beta\in J_1^\alpha} \int_{B_\varepsilon^\eta(I_3^\beta)} |\nabla u|^2\,dx + \frac{c}{\eta^2}\sum_{\beta\in J_1^\alpha} \int_{Q_\eta^\alpha} |w_\eta^\beta - w_\eta^\alpha|^2\,dx\,. \qquad (5.31)\end{aligned}$$

Since $Q_\eta^\alpha \cap B_\varepsilon^\alpha \cap B_\varepsilon^\beta \neq \emptyset$ for every $\beta \in J_1^\alpha$, by Poincaré's inequality we have

$$\int_{Q_\eta^\alpha} |w_\eta^\beta - w_\eta^\alpha|^2\,dx \le c\,\eta^2 \int_{Q_\eta^\alpha} |\nabla w_\eta^\beta - \nabla w_\eta^\alpha|^2\,dx\,. \qquad (5.32)$$

From (5.30) it follows that for every $\alpha \in I$ and for every $\beta \in J_1^\alpha$

$$\int_{Q_\eta^\alpha} |w_\eta^\beta - w_\eta^\alpha|^2\,dx \le c\,\eta^2 \int_{B_\varepsilon^\eta(I_3^\beta)} |\nabla u|^2\,dx + c\,\eta^2 \int_{B_\varepsilon^\eta(I_3^\alpha)} |\nabla u|^2\,dx\,,$$

so that by (5.31) for every $\alpha \in I$ we have

$$\int_{Q_\eta^\alpha} |\nabla w|^2\,dx \le c \int_{B_\varepsilon^\eta(I_4^\alpha)} |\nabla u|^2\,dx\,,$$

where $I_4^\alpha = \{\beta \in I : |\beta - \alpha|_\infty \le 4\}$. By taking the sum for $\alpha \in I$ we obtain (5.27).

If $u \in L^\infty(B_\varepsilon^\eta(I))$, first we construct $w$ as above, and then we take $(w \wedge k) \vee (-k)$, with $k = \|u\|_{L^\infty(B_\varepsilon^\eta(I))}$. □

**Proof of Proposition 5.5** It is not a restriction to suppose $a = 1$. Fix $\varepsilon > 0$, $\delta > 0$, and $\xi \in S^{n-1}$. We may assume that $\delta < 1/2$ and that $\xi = e_1 =$

$(1,0,\dots,0)$. Let $\eta = 2(1-\delta)\varepsilon$ and let $I_\rho$, $B^\alpha_\rho$, $Q^\alpha_\eta$, $Q_\eta(I)$, $I^r_\rho(t)$, $J^r_\rho(t)$, and $B^\eta_\varepsilon(I)$ be the sets defined in (5.11), (5.12), (5.15), and (5.21). As in the proof of Proposition 5.2 we obtain

$$F_\varepsilon(u,A) \geq \frac{1}{\varepsilon}\sum_{\alpha\in I_\sigma} \eta^n f\Big(\varepsilon \fint_{B^\alpha_\varepsilon} |\nabla u|^2\,dx\Big). \tag{5.33}$$

As $1/2 < \varepsilon/\eta = 2^{-1}(1-\delta)^{-1/2} < \sqrt{2}/2$, we can apply Lemma 5.6 to $I = I^\varepsilon_\sigma(b)$ and $s = \varepsilon/\eta$. In this way we obtain a function $w \in SBV(Q_\eta(I^\varepsilon_\sigma(b)))$ such that $w = u$ in $Q_\eta(I^\varepsilon_\sigma(b)) \cap B^\eta_\varepsilon(I^\varepsilon_\sigma(b))$, $\|w\|_{L^\infty(Q_\eta(I^\varepsilon_\sigma(b)))} \leq \|u\|_{L^\infty(A)}$,

$$S(w) \subseteq \bigcup_{\alpha\in b_3 I^\varepsilon_\sigma(b)} \partial_1 Q^\alpha_\eta \qquad \text{up to an } \mathcal{H}^{n-1}\text{-negligible set}, \tag{5.34}$$

and

$$\int_{Q_\eta(I^\varepsilon_\sigma(b))} |\nabla w|^2\,dx \leq c\int_{B^\eta_\varepsilon(I^\varepsilon_\sigma(b))} |\nabla u|^2\,dx\,, \tag{5.35}$$

where $c$ is a constant depending only on $n$ and $\delta$.

We now define $w$ in $A \setminus Q_\eta(I^\varepsilon_\sigma(b))$ as follows. Let $\alpha \in \mathbf{Z}^n \setminus I^\varepsilon_\sigma(b)$, let $Q^{\alpha+}_\eta = Q^\alpha_\eta \cap \{x^1 > \eta\alpha^1\}$, and let $\beta = \alpha + e_1$. If $\beta \in I^\varepsilon_\sigma(b)$, we define $w$ on $Q^{\alpha+}_\eta \cap A$ by a reflection of $w\big|_{Q^\beta_\eta}$ with respect to the common face of $Q^\alpha_\eta$ and $Q^\beta_\eta$; more precisely we set

$$w(x^1,\dots,x^n) = w(2\eta\alpha^1 + \eta - x^1, x^2,\dots,x^n)$$

for $x = (x^1,\dots,x^n) \in Q^{\alpha+}_\eta \cap A$. If $\beta \notin I^\varepsilon_\sigma(b)$, then we set $w = 0$ on $Q^{\alpha+}_\eta \cap A$. Symmetrically, we define $w$ on $Q^\alpha_\eta \cap \{x^1 < \eta\alpha^1\} \cap A$, according to the cases $\alpha - e_1 \in I^\varepsilon_\sigma(b)$ or not.

By construction $w$ belongs to $SBV(A) \cap L^\infty(A)$ and $\|w\|_{L^\infty(A)} \leq \|u\|_{L^\infty(A)}$. By (5.35) we have

$$\int_A |\nabla w|^2\,dx \leq 2\int_{Q_\eta(I^\varepsilon_\sigma(b))} |\nabla w|^2\,dx \leq 2c\int_A |\nabla u|^2\,dx\,.$$

Let $F^\alpha_\eta = Q^\alpha_\eta \cap \{x^1 = \eta\alpha^1\}$. Since $w \in H^1(Q^\alpha_\eta)$ for every $\alpha \in I^\varepsilon_\sigma(b)$, by (5.34) we have

$$S(w) \cap Q_\eta(I_\sigma) \subseteq \Big(\bigcup_{\alpha\in b_3 I^\varepsilon_\sigma(b)} \partial_1 Q^\alpha_\eta\Big) \cup \Big(\bigcup_{\alpha\in J^\varepsilon_\sigma(b)} \partial_1 Q^\alpha_\eta\Big) \cup \Big(\bigcup_{\alpha\in J^\varepsilon_\sigma(b)} F^\alpha_\eta\Big) \tag{5.36}$$

up to an $\mathcal{H}^{n-1}$-negligible set.

In order to estimate $\mathcal{H}^{n-1}(S(w) \cap A_{8\sigma})$ we observe that $A_{8\sigma} \subseteq Q_\eta(I_{7\sigma})$. We now prove that

$$I_{7\sigma} \cap b_3 I^\varepsilon_\sigma(b) \subseteq e_3 J^\varepsilon_\sigma(b)\,. \tag{5.37}$$

If $\alpha \in I_{7\sigma} \cap b_3 I^\varepsilon_\sigma(b)$, by (5.20) there exists $\beta \in \mathbf{Z}^n \setminus I^\varepsilon_\sigma(b)$ such that $|\beta - \alpha|_\infty \leq 3$; hence, $|\eta\beta - \eta\alpha| \leq 3\eta\sqrt{n} \leq 6\sigma$. As $\alpha \in I_{7\sigma}$, we have $\beta \in I_\sigma$. Since $\beta \notin I^\varepsilon_\sigma(b)$, we also have $\beta \in J^\varepsilon_\sigma(b)$, hence $\alpha \in e_3 J^\varepsilon_\sigma(b)$ and (5.37) is proved. By (5.36) and (5.37) we have

$$S(w) \cap A_{8\sigma} \subseteq S(w) \cap Q_\eta(I_{7\sigma}) \subseteq \Big( \bigcup_{\alpha \in e_3 J^\varepsilon_\sigma(b)} \partial_1 Q^\alpha_\eta \Big) \cup \Big( \bigcup_{\alpha \in J^\varepsilon_\sigma(b)} F^\alpha_\eta \Big) \tag{5.38}$$

up to an $\mathcal{H}^{n-1}$-negligible set. This implies that

$$\mathcal{H}^{n-1}(S(w) \cap A_{8\sigma}) \leq 2n\, \eta^{n-1} \#(e_3 J^\varepsilon_\sigma(b)) \leq 2n 7^n\, \eta^{n-1} \# J^\varepsilon_\sigma(b)\,. \tag{5.39}$$

By (5.33) we have

$$b\, \eta^n \# J^\varepsilon_\sigma(b) \leq \varepsilon\, F_\varepsilon(u, A)\,, \tag{5.40}$$

so that (5.39) gives

$$\mathcal{H}^{n-1}(S(w) \cap A_{8\sigma}) \leq \frac{2n7^n}{b} \frac{\varepsilon}{\eta} F_\varepsilon(u, A) = \frac{n7^n}{(1-\delta)\, b} F_\varepsilon(u, A)\,.$$

As $\langle \nu_w, e_1 \rangle = 0$ in $S(w) \cap \partial_1 Q^\alpha_\eta$, from (5.38) and (5.40) we have

$$\begin{aligned} &2\,(1-\delta)\, b \int_{S(w)\cap A_{8\sigma}} |\langle \nu_w, e_1 \rangle|\, d\mathcal{H}^{n-1} \\ &\leq 2\,(1-\delta)\, b \sum_{\alpha \in J^\varepsilon_\sigma(b)} \int_{S(w)\cap F^\alpha_\eta} |\langle \nu_w, e_1 \rangle|\, d\mathcal{H}^{n-1} \\ &\leq 2\,(1-\delta)\, b\, \eta^{n-1} \# J^\varepsilon_\sigma(b) \leq 2\,(1-\delta) \frac{\varepsilon}{\eta} F_\varepsilon(u, A) = F_\varepsilon(u, A)\,. \end{aligned}$$

Let $E = Q_\eta(I_\sigma) \cap B^\eta_\varepsilon(I_\sigma)$. For every $x \in A_{2\sigma}$ there exists $\alpha \in I_\sigma$ such that $|x - \eta\alpha| \leq \sigma$. This implies that $E \cap B_{2\sigma}(x) \supseteq Q^\alpha_\eta \cap B^\alpha_\varepsilon \supseteq B^\alpha_{\varepsilon/2}$, hence $|E \cap B_{2\sigma}(x)| \geq |B_{\varepsilon/2}|$. As $w = u$ in $Q_\eta(I^\varepsilon_\sigma(b)) \cap B^\eta_\varepsilon(I^\varepsilon_\sigma(b))$, by (5.40) we have

$$\|w - u\|_{L^1(E)} \leq \eta^n \# J^\varepsilon_\sigma(b)\, \|w - u\|_{L^\infty(A)} \leq \frac{2\varepsilon}{b} F_\varepsilon(u, A)\, \|u\|_{L^\infty(A)}\,,$$

which concludes the proof of the proposition. □

### 5.1.3 *Estimate from below of the Γ-limit*

With fixed a sequence of positive real numbers $(\varepsilon_j)$ converging to 0 as $j \to \infty$, for every $u \in L^1(\Omega)$ and for every open set $A \subseteq \Omega$ we define

$$F'(u, A) = \Gamma\text{-}\liminf_j F_{\varepsilon_j}(u, A)\,.$$

**Remark 5.8** For every $u \in L^1(\Omega)$ and for every $k > 0$ let $u_k = (u \wedge k) \vee (-k)$. Since $F_\varepsilon(u, A) \geq F_\varepsilon(u_k, A)$, we obtain $F'(u, A) \geq F'(u_k, A)$ for every $u \in L^1(\Omega)$, for every open set $A \subseteq \Omega$, and for every $k > 0$.

We start with the following estimates from below of $F'(u, A)$, where the volume integral $\int_A |\nabla u|^2\,dx$ and the surface integral $\int_{S(u)\cap A} |\langle \nu_u, \xi\rangle|\,d\mathcal{H}^{n-1}$ are considered separately.

**Proposition 5.9** *Let $f(t) = \min\{at, b\}$. Let $u \in L^\infty(\Omega)$ and let $A$ be an open subset of $\Omega$ such that $F'(u, A) < +\infty$. Then $u \in SBV(A)$ and*

$$\mathcal{H}^{n-1}(S(u) \cap A) < +\infty\,, \tag{5.41}$$

$$F'(u, A) \geq a \int_A |\nabla u|^2\,dx\,, \tag{5.42}$$

$$F'(u, A) \geq 2b \int_{S(u)\cap A} |\langle \nu_u, \xi\rangle|\,d\mathcal{H}^{n-1} \tag{5.43}$$

*for every $\xi \in S^{n-1}$.*

To prove Proposition 5.9 we need the following lemma.

**Lemma 5.10** *Let $A$ and $A'$ be two bounded open subsets of $\mathbf{R}^n$ with $A' \subset\subset A$, and let $(E_j)$ be a sequence of Borel subsets of $A$ such that $(\chi_{E_j})$ converges to $\psi$ weakly* in $L^\infty(A)$. Suppose that there exists a sequence of real numbers $(\sigma_j)$ converging to $0$ as $j \to 0$ such that for every $x \in A'$*

$$|E_j \cap B_{\sigma_j}(x)| \geq c\,\sigma_j^n\,, \tag{5.44}$$

*with a constant $c$ independent of $x$ and $j$. Then $\psi \geq c2^{-n}$ almost everywhere in $A'$.*

**Proof** Let $Q$ be a cube of side $r$ contained in $A'$. Then $Q$ contains at least $[r/(2\varepsilon_j)]^n$ disjoint balls of radius $\varepsilon_j$, where $[t]$ denotes the integer part of $t$. By (5.44) this implies that

$$\liminf_j |E_j \cap Q| \geq c\,r^n/2^n\,;$$

hence, $\int_Q \psi\,dx \geq c/2^n\,|Q|$. As this inequality holds for every cube $Q$ contained in $A'$, we conclude that $\psi \geq c/2^n$ almost everywhere in $A'$. □

**Proof of Proposition 5.9** By the definition of Γ-liminf there exists a sequence $(u_j)$ converging to $u$ in $L^1(\Omega)$ such that

$$F'(u, A) = \liminf_j F_{\varepsilon_j}(u_j, A)\,. \tag{5.45}$$

Replacing, if necessary, $u_j$ by $(u_j \wedge k) \vee (-k)$, with $k = \|u\|_{L^\infty(\Omega)}$, it is not restrictive to suppose that $\|u_j\|_{L^\infty(\Omega)} \leq \|u\|_{L^\infty(\Omega)}$ for all $j$. With fixed $\delta > 0$ and an open set $A' \subset\subset A$, we apply Proposition 5.2 to the function $u_j$. We obtain a function $v_j \in SBV(A')$ such that for $j$ large enough

$$
\begin{aligned}
&(1-\delta)a\int_{A'} |\nabla v_j|^2\,dx \le F_{\varepsilon_j}(u_j, A)\,,\\
&\mathcal{H}^{n-1}(S(v_j)\cap A') \le c\,F_{\varepsilon_j}(u_j, A)\,,\\
&\|v_j\|_{L^\infty(A')} \le \|u\|_{L^\infty(\Omega)}\,,\\
&\|v_j - u_j\|_{L^1(A')} \le c\,\varepsilon_j\,F_{\varepsilon_j}(u_j, A)\,\|u\|_{L^\infty(\Omega)}\,,
\end{aligned}
\tag{5.46}
$$

where $c$ is a constant independent of $j$. As the sequence $\big(F_{\varepsilon_j}(u_j, A)\big)$ is bounded, we conclude that $(v_j)$ converges to $u$ in $L^1(A')$, so that by Theorem 2.3 we have $u \in SBV(A')$ and

$$
\begin{aligned}
&\int_{A'} |\nabla u|^2\,dx \le \liminf_j \int_{A'} |\nabla v_j|^2\,dx\,,\\
&\mathcal{H}^{n-1}(S(u)\cap A') \le \liminf_j \mathcal{H}^{n-1}(S(v_j)\cap A')\,.
\end{aligned}
$$

By (5.46) and (5.45) this implies

$$
\begin{aligned}
&(1-\delta)a\int_{A'} |\nabla u|^2\,dx \le \liminf_j F_{\varepsilon_j}(u_j, A) = F'(u, A)\\
&\mathcal{H}^{n-1}(S(u)\cap A') \le c\liminf_j F_{\varepsilon_j}(u_j, A) = c\,F'(u, A)\,.
\end{aligned}
$$

By taking the limit as $\delta \to 0$ and as $A'$ converges increasingly to $A$ we obtain that $u \in SBV(A)$ and that (5.41) and (5.42) hold.

We now prove (5.43). Passing to a subsequence, we may assume that the lower limit in (5.45) is replaced by a limit. Passing to a further subsequence, by Proposition 5.5 and by Lemma 5.10 there exist a sequence $(w_j)$ in $SBV(A')$, a sequence $(E_j)$ of Borel subsets of $A'$, and a function $\psi \in L^\infty(A')$ with $\psi(x) > 0$ for almost every $x \in A'$, such that $(\chi_{E_j})$ converges to $\psi$ weakly* in $L^\infty(A')$ and for $j$ large enough

$$
\begin{aligned}
&2(1-\delta)\,b\int_{S(w_j)\cap A'} |\langle \nu_{w_j}, \xi\rangle|\,d\mathcal{H}^{n-1} \le F_{\varepsilon_j}(u_j, A)\,,\\
&\int_{A'} |\nabla w_j|^2\,dx \le c\,F_{\varepsilon_j}(u_j, A)\,,\\
&\mathcal{H}^{n-1}(S(w_j)\cap A') \le F_{\varepsilon_j}(u_j, A)\,,\\
&\|w_j\|_{L^\infty(A')} \le \|u\|_{L^\infty(\Omega)}\,,\\
&\|w_j - u_j\|_{L^1(E_j)} \le c\,\varepsilon_j\,F_{\varepsilon_j}(u_j, A)\,\|u\|_{L^\infty(\Omega)}\,,
\end{aligned}
\tag{5.47}
$$

where $c$ is a constant independent of $j$. By Theorem 2.3 there exists a subsequence (not relabelled) and a function $w \in SBV(A')$ such that $(w_j)$ converges to $w$ in $L^1(A')$ and

$$\int_{S(w)\cap A'} |\langle \nu_w, \xi\rangle|\, d\mathcal{H}^{n-1} \le \liminf_j \int_{S(w_j)\cap A'} |\langle \nu_{w_j}, \xi\rangle|\, d\mathcal{H}^{n-1}. \tag{5.48}$$

Since the sequence $(F_{\varepsilon_j}(u_j, A))$ is bounded, the last inequality in (5.47) implies that $((w_j - u)\chi_{E_j})$ converges to 0 strongly in $L^1(A')$. As $(w_j - u)$ converges to $w - u$ strongly in $L^1(A')$ and $(\chi_{E_j})$ converges to $\psi$ weakly* in $L^\infty(A')$, we also have that $((w_j - u)\chi_{E_j})$ converges to $(w-u)\psi$ weakly in $L^1(A')$. Therefore $(w-u)\psi = 0$ almost everywhere in $A'$. Since $\psi > 0$, we conclude that $w = u$ almost everywhere in $A'$. By (5.48), (5.47), and (5.45) this implies that

$$2\,(1-\delta)\,b \int_{S(u)\cap A'} |\langle \nu_u, \xi\rangle|\, d\mathcal{H}^{n-1} \le \liminf_j F_{\varepsilon_j}(u_j, A) = F'(u, A).$$

Taking the limit as $\delta \to 0$ and as $A'$ converges increasingly to $A$ we obtain (5.43). □

We now can prove the complete estimate from below for $F'(u, A)$, using Proposition 1.16.

**Proposition 5.11** *Let $u \in L^\infty(\Omega)$ and let $A$ be an open subset of $\Omega$ such that $F'(u, A) < +\infty$. Then $u \in SBV(A)$ and*

$$F'(u, A) \ge a \int_A |\nabla u|^2\, dx + 2b\mathcal{H}^{n-1}(S(u)\cap A).$$

**Proof** Given $u \in L^1(\Omega)$, the set functions $A \mapsto F_\varepsilon(u, A)$ are increasing and superadditive. Consequently also the set function $A \mapsto F'(u, A)$ is increasing and superadditive.

First, let $f(t) = \min\{at, b\}$. We already know from Proposition 5.9 that $u \in SBV(A)$ and that $\mathcal{H}^{n-1}(S(u)\cap A) < +\infty$. The thesis is immediately obtained by applying Proposition 1.16 with $\mu(A) = F'(u, A)$, $\lambda = \mathcal{L}_n + \mathcal{H}^{n-1} \llcorner S(u)$,

$$\psi_0(x) = \begin{cases} |\nabla u|^2 & \text{if } x \notin S(u) \\ 0 & \text{if } x \in S(u) \end{cases}$$

and

$$\psi_i(x) = \begin{cases} 0 & \text{if } x \notin S(u) \\ |\langle \xi_i, \nu_u\rangle| & \text{if } x \in S(u), \end{cases}$$

where $(\xi_i)$ $(i = 1, 2, \ldots)$ is a dense sequence in $S^{n-1}$.

For a general $f$, we apply the previous step with functions $f_i(t) = \min\{a_i t, b_i\}$ such that $f_i \le f$, $\sup_i a_i = a$ and $\sup_i b_i = b$, and then use Proposition 1.16 as in Proposition 3.25. □

**Proposition 5.12** *Let $u \in L^1(\Omega)$ and let $A$ be an open subset of $\Omega$ such that $F'(u, A) < +\infty$. Then $u \in GSBV(A)$ and*

$$F'(u, A) \geq a \int_A |\nabla u|^2 \, dx + 2b\mathcal{H}^{n-1}(S(u) \cap A).$$

**Proof** For every integer $k > 0$ we consider the truncation $u_k = (u \wedge k) \vee (-k)$. By Remark 5.8 and by Proposition 5.11 we have $u_k \in SBV(A)$ and

$$F'(u, A) \geq F'(u_k, A) \geq a \int_A |\nabla u_k|^2 \, dx + 2b\mathcal{H}^{n-1}(S(u_k) \cap A).$$

The conclusion follows by taking the limit as $k \to \infty$. □

5.1.4 *Estimate from above of the* $\Gamma$*-limit*

As in the previous section, fix a sequence of positive real numbers $(\varepsilon_j)$ which converges to 0 as $j \to \infty$. For every $u \in L^1(\Omega)$ we define

$$F''(u) = \Gamma\text{-}\limsup_j F_{\varepsilon_j}(u).$$

In order to obtain an estimate from above of $F''(u)$ we will use Lemma 4.11.

**Proposition 5.13** *For every $u \in GSBV(\Omega) \cap L^1(\Omega)$ we have*

$$F''(u) \leq a \int_\Omega |\nabla u|^2 \, dx + 2b\mathcal{H}^{n-1}(S(u)). \tag{5.49}$$

**Proof** Let $\Omega'$ be a bounded open subset of $\mathbf{R}^n$ with $\Omega \subset\subset \Omega'$. By Lemma 4.11 it is sufficient to prove that

$$F''(u) \leq a \int_\Omega |\nabla u|^2 \, dx + 2b\mathcal{H}^{n-1}(S(u) \cap \overline{\Omega}) \tag{5.50}$$

when $u \in SBV(\Omega') \cap L^\infty(\Omega')$, $\mathcal{H}^{n-1}(S(u)) < +\infty$, $\mathcal{H}^{n-1}(\overline{S}(u) \setminus S(u)) = 0$, $\mathcal{H}^{n-1}(S(u) \cap K) = \mathcal{M}^{n-1}(S(u) \cap K)$ for every compact set $K \subset\subset \Omega'$, and $\nabla u \in L^2(\Omega'; \mathbf{R}^n)$. Note that in this case $u \in H^1(\Omega' \setminus \overline{S}(u))$.

For every $\varepsilon > 0$ we choose a function $u_\varepsilon \in H^1(\Omega) \cap L^\infty(\Omega)$ such that $u_\varepsilon = u$ in the set $\{x \in \Omega : \operatorname{dist}(x, S(u) \cap \overline{\Omega}) > \varepsilon^2\}$ and $\|u_\varepsilon\|_{L^\infty(\Omega)} \leq \|u\|_{L^\infty(\Omega)}$. As $\mathcal{M}^{n-1}(S(u) \cap \overline{\Omega}) < +\infty$, $(u_\varepsilon)$ converges to $u$ in $L^1(\Omega)$ as $\varepsilon \to 0$. Note that $u_\varepsilon = u$ in $B_\varepsilon(x) \cap \Omega$ if $\operatorname{dist}(x, S(u) \cap \overline{\Omega}) \geq \varepsilon + \varepsilon^2$. Hence,

$$F_\varepsilon(u_\varepsilon) \leq \frac{1}{\varepsilon} \int_\Omega f\Big(\varepsilon \fint_{B_\varepsilon(x) \cap \Omega} |\nabla u(y)|^2 \, dy\Big) \, dx + \frac{b}{\varepsilon} |V_\varepsilon|, \tag{5.51}$$

where $V_\varepsilon = \{x \in \Omega : \operatorname{dist}(x, S(u) \cap \overline{\Omega}) < \varepsilon + \varepsilon^2\}$.

For every $x \in \Omega$ and for every $\varepsilon > 0$ we set

$$g_\varepsilon(x) = \fint_{B_\varepsilon(x)\cap\Omega} |\nabla u(y)|^2\,dy \qquad \text{and} \qquad g(x) = |\nabla u(x)|^2\,. \tag{5.52}$$

Since $g \in L^1(\Omega)$, by the Lebesgue Differentiation Theorem $(g_\varepsilon)$ converges to $g$ in $L^1(\Omega)$ and almost everywhere in $\Omega$.

As $f$ is non-decreasing, by (5.1) there exists $c \geq a$ such that $f(t) \leq ct$ for every $t \geq 0$. Therefore $\frac{1}{\varepsilon} f(\varepsilon g_\varepsilon(x)) \leq c\, g_\varepsilon(x)$ for every $x \in \Omega$ and for every $\varepsilon > 0$. By (5.1) for almost every $x \in \Omega$ we have

$$\lim_{\varepsilon\to 0} \frac{1}{\varepsilon} f(\varepsilon g_\varepsilon(x)) = a g(x)\,,$$

so that by the Dominated Convergence Theorem

$$\lim_{\varepsilon\to 0} \frac{1}{\varepsilon} \int_\Omega f(\varepsilon g_\varepsilon(x))\,dx = a \int_\Omega g(x)\,dx\,. \tag{5.53}$$

From (5.51), (5.52), and (5.53) we obtain

$$F''(u) \leq \limsup_j F_{\varepsilon_j}(u_{\varepsilon_j}) \leq a \int_\Omega |\nabla u|^2\,dx \,+\, 2\,b\mathcal{M}^{n-1}(S(u)\cap\overline{\Omega})\,,$$

which proves (5.50) and concludes the proof of the proposition. □

**Remark 5.14** Theorem 5.1 holds also if we modify the functionals $F_\varepsilon$ by setting $F_\varepsilon(u) = \int_\Omega f(\varepsilon w_u(x))\,dx$ for all $u \in L^1(\Omega)$, where

$$w_u(x) = \begin{cases} \displaystyle\fint_{B_\varepsilon(x)\cap\Omega} |\nabla u|^2\,dy & \text{if } u \in H^1(B_\varepsilon(x)\cap\Omega) \\ +\infty & \text{otherwise,} \end{cases}$$

and $f(+\infty) = b$. It suffices that the proof of the lower inequalities holds unchanged. We also remark that in this case we can take $u_\varepsilon = u$ as a recovery sequence in Proposition 5.13.

### 5.1.5 *Some convergence results*

In this section we obtain approximate solutions to problems involving the Mumford-Shah functional. Before stating the convergence results we prove the following lemma, where we use the notation introduced in (5.5).

**Lemma 5.15** *There exists a constant $C_\Omega$, with $0 < C_\Omega < 1/2$, such that*

$$|B_\rho(x)\cap\Omega| \geq C_\Omega\,|B_\rho| \tag{5.54}$$

*for every $x \in \Omega$ and for every $\rho > 0$. Moreover*

$$\int_{\Omega_\varepsilon} |\nabla u(x)|^2\,dx \,+\, C_\Omega \int_{\Omega\setminus\Omega_\varepsilon} |\nabla u(x)|^2\,dx \leq \int_\Omega \Big(\frac{1}{|B_\varepsilon|} \int_{B_\varepsilon(x)\cap\Omega} |\nabla u(y)|^2\,dy\Big)\,dx$$

$$\leq \int_\Omega \Big( \fint_{B_\varepsilon(x)\cap\Omega} |\nabla u(y)|^2\,dy \Big)\,dx \leq \int_{\Omega_{2\varepsilon}} |\nabla u(x)|^2\,dx + \frac{1}{C_\Omega}\int_{\Omega\setminus\Omega_{2\varepsilon}} |\nabla u(x)|^2\,dx$$

*for every $u \in H^1(\Omega)$ and for every $\varepsilon > 0$.*

**Proof** Inequality (5.54) follows easily from the fact that $\Omega$ is bounded and has a Lipschitz boundary. Since $\chi_{B_\varepsilon(x)}(y) = \chi_{B_\varepsilon(y)}(x)$, for every $\varepsilon > 0$ we have

$$\begin{aligned}
\int_\Omega \Big( \fint_{B_\varepsilon(x)\cap\Omega} |\nabla u(y)|^2\,dy \Big)\,dx &= \int_\Omega \Big( \int_\Omega \frac{\chi_{B_\varepsilon(x)}(y)}{|B_\varepsilon(x)\cap\Omega|} |\nabla u(y)|^2\,dy \Big)\,dx \\
&= \int_\Omega |\nabla u(y)|^2 \Big( \int_\Omega \frac{\chi_{B_\varepsilon(y)}(x)}{|B_\varepsilon(x)\cap\Omega|}\,dx \Big)\,dy \\
&\leq \int_{\Omega_{2\varepsilon}} |\nabla u(y)|^2\,dy + \frac{1}{C_\Omega}\int_{\Omega\setminus\Omega_{2\varepsilon}} |\nabla u(y)|^2\,dy\,.
\end{aligned}$$

Similarly we have

$$\begin{aligned}
\int_\Omega \Big( \frac{1}{|B_\varepsilon|} \int_{B_\varepsilon(x)\cap\Omega} |\nabla u(y)|^2\,dy \Big)\,dx &= \int_\Omega \Big( \int_\Omega \frac{\chi_{B_\varepsilon(y)}(x)}{|B_\varepsilon|} |\nabla u(y)|^2\,dy \Big)\,dx \\
&= \int_\Omega |\nabla u(y)|^2 \frac{|B_\varepsilon(y)\cap\Omega|}{|B_\varepsilon|}\,dy \\
&\geq \int_{\Omega_\varepsilon} |\nabla u(y)|^2\,dy + C_\Omega \int_{\Omega\setminus\Omega_\varepsilon} |\nabla u(y)|^2\,dy\,,
\end{aligned}$$

which concludes the proof of the lemma. □

We now state a simple variant of Theorem 5.1. Let $(f_\varepsilon)_{\varepsilon>0}$ be a family of Borel functions from $[0,+\infty)$ to $[0,+\infty)$ and, for every $\varepsilon > 0$, let $\mathcal{F}_\varepsilon$ be the functional defined as in (5.2), with $f$ replaced by $f_\varepsilon$.

**Corollary 5.16** *Assume that there exist a family $(a_\varepsilon)_{\varepsilon>0}$ of real numbers, with*

$$\lim_{\varepsilon\to 0} \frac{a_\varepsilon}{\varepsilon} = 0\,, \tag{5.55}$$

*and a non-decreasing continuous function $f : [0,+\infty) \to [0,+\infty)$ satisfying (5.1), such that*

$$f(t) \leq f_\varepsilon(t) \leq f(t) + a_\varepsilon t \tag{5.56}$$

*for every $\varepsilon > 0$ and for every $t \geq 0$. Then $(\mathcal{F}_\varepsilon)$ $\Gamma$-converges in $L^1(\Omega)$ to the functional $F$ defined by (5.3).*

**Proof** Fix a sequence of positive real numbers $(\varepsilon_j)$ converging to 0 as $j \to \infty$. For every $u \in L^1(\Omega)$ we define

$$\mathcal{F}'(u) = \Gamma\text{-}\liminf_j \mathcal{F}_{\varepsilon_j}(u) \qquad \text{and} \qquad \mathcal{F}''(u) = \Gamma\text{-}\limsup_j \mathcal{F}_{\varepsilon_j}(u)\,.$$

By (5.56) we have

$$F_\varepsilon(u) \le \mathcal{F}_\varepsilon(u) \le F_\varepsilon(u) + a_\varepsilon \int_\Omega \Big( \fint_{B_\varepsilon(x)\cap\Omega} |\nabla u(y)|^2 \, dy \Big) \, dx \tag{5.57}$$

for every $u \in H^1(\Omega) \cap L^1(\Omega)$. As $(F_{\varepsilon_j})$ $\Gamma$-converges to $F$ (Theorem 5.1), (5.57) implies that $F(u) \le \mathcal{F}'(u)$.

It remains to prove that $\mathcal{F}''(u) \le F(u)$. Let $\Omega'$ be a bounded open subset of $\mathbf{R}^n$ with $\Omega \subset\subset \Omega'$. As in the proof of Proposition 5.13 it is enough to prove that

$$\mathcal{F}''(u) \le a \int_\Omega |\nabla u(x)|^2 \, dx + 2b\mathcal{H}^{n-1}(S(u) \cap \overline{\Omega}) \tag{5.58}$$

when $u \in SBV(\Omega') \cap L^\infty(\Omega')$, $\mathcal{H}^{n-1}(S(u)) < +\infty$, $\mathcal{H}^{n-1}(\overline{S}(u) \setminus S(u)) = 0$, $\mathcal{H}^{n-1}(S(u) \cap K) = \mathcal{M}^{n-1}(S(u) \cap K)$, and $\nabla u \in L^2(\Omega'; \mathbf{R}^n)$ for every compact set $K \subseteq \Omega$. Note that in this case $u \in H^1(\Omega' \setminus \overline{S}(u))$.

For every $\varepsilon > 0$ let $\rho_\varepsilon = \varepsilon^{3/4} a_\varepsilon^{1/4}$, so that by (5.55)

$$\lim_{\varepsilon \to 0} \frac{\rho_\varepsilon}{\varepsilon} = \lim_{\varepsilon \to 0} \frac{\varepsilon \, a_\varepsilon}{\rho_\varepsilon^2} = 0 \,. \tag{5.59}$$

For $\rho_\varepsilon < \mathrm{dist}\,(\Omega, \partial\Omega')$ we consider the convolution $u_\varepsilon = u * \psi_{\rho_\varepsilon}$, where $\psi \in C_c^\infty(B_1)$ is a mollifier, and $\psi_\rho(x) = \rho^{-n}\psi(x/\rho)$. Note that $\|\nabla \psi_\rho\|_{L^\infty(\mathbf{R}^n;\mathbf{R}^n)} \le k/\rho^{n+1}$ for a suitable constant $k$, so that for every $x \in \Omega$ we have

$$|\nabla u_\varepsilon(x)| \le \int_{B_{\rho_\varepsilon}(x)} |u(y)| \, |\nabla \psi_{\rho_\varepsilon}(x-y)| \, dy \le \frac{c}{\rho_\varepsilon} \|u\|_{L^\infty(\Omega')} \,, \tag{5.60}$$

where $c$ is a constant independent of $\varepsilon$ and $x$. Let

$$M_\varepsilon = \{x \in \Omega : \mathrm{dist}\,(x, S(u)) < \varepsilon + \rho_\varepsilon\}, \quad R_\varepsilon = \{x \in \Omega : \mathrm{dist}\,(x, S(u)) \ge \varepsilon + \rho_\varepsilon\}.$$

For every $x \in \Omega$ and for every $\varepsilon > 0$ with $\varepsilon + \rho_\varepsilon < \mathrm{dist}\,(\Omega, \partial\Omega')$ we set

$$g_\varepsilon(x) = \fint_{B_\varepsilon(x)\cap\Omega} |\nabla u_\varepsilon(y)|^2 \, dy \,, \qquad h_\varepsilon(x) = \Big(\frac{\varepsilon + \rho_\varepsilon}{\varepsilon}\Big)^n \fint_{B_{\varepsilon+\rho_\varepsilon}(x)} |\nabla u(y)|^2 \, dy \,, \tag{5.61}$$

and $g(x) = |\nabla u(x)|^2$. Since $g \in L^1(\Omega')$, by (5.59) by the Lebesgue Differentiation Theorem $(h_\varepsilon)$ converges to $g$ in $L^1(\Omega)$ and almost everywhere in $\Omega$. For every $x \in R_\varepsilon$ we have $u \in H^1\big(B_{\varepsilon+\rho_\varepsilon}(x)\big)$. By Jensen's inequality this implies that for every $x \in R_\varepsilon$ we have

$$g_\varepsilon(x) \le \frac{1}{C_\Omega} h_\varepsilon(x) \,, \tag{5.62}$$

where $C_\Omega$ is the constant which appears in (5.54). If $x \in R_\varepsilon$ and $\mathrm{dist}\,(x, \partial\Omega) > \varepsilon$, then we also have

$$g_\varepsilon(x) \le h_\varepsilon(x) \,. \tag{5.63}$$

Since $\big(h_\varepsilon(x)\big)$ converges to $g(x)$ for almost every $x \in \Omega$, from (5.1) and (5.63) we obtain

$$\limsup_{\varepsilon\to 0} \frac{1}{\varepsilon} f(\varepsilon g_\varepsilon(x)) \le a g(x) \tag{5.64}$$

for almost every $x \in \Omega$. As $f$ is non-decreasing, by (5.1) there exists a constant $c \ge a$ such that $f(t) \le c\,t$ for every $t \ge 0$, so that (5.62) gives

$$\frac{1}{\varepsilon} f(\varepsilon g_\varepsilon(x)) \le \frac{c}{C_\Omega} h_\varepsilon(x) \tag{5.65}$$

for every $x \in R_\varepsilon$. Since $(h_\varepsilon)$ converges to $g$ in $L^1(\Omega)$ and almost everywhere in $\Omega$, inequalities (5.64) and (5.65) imply, by Fatou's Lemma, that

$$\limsup_{\varepsilon\to 0} \frac{1}{\varepsilon} \int_{R_\varepsilon} f(\varepsilon g_\varepsilon(x))\, dx \le a \int_\Omega g(x)\, dx\,. \tag{5.66}$$

Since $f(t) \le b$ for every $t \ge 0$, we have

$$F_\varepsilon(u_\varepsilon) \le \frac{1}{\varepsilon} \int_{R_\varepsilon} f(\varepsilon g_\varepsilon(x))\, dx + \frac{b}{\varepsilon}|M_\varepsilon|\,. \tag{5.67}$$

Let $K$ be a compact subset of $\Omega'$ whose interior contains $\overline{\Omega}$. Since $M_\varepsilon \subseteq \{x \in \mathbf{R}^n : \mathrm{dist}\,(x, S(u)\cap K) < \varepsilon + \rho_\varepsilon\}$ for $\varepsilon + \rho_\varepsilon < \mathrm{dist}\,(\Omega, \partial K)$, from (5.59) we obtain

$$\limsup_{\varepsilon\to 0} \frac{|M_\varepsilon|}{\varepsilon} \le \mathcal{M}^{n-1}(S(u)\cap K) = \mathcal{H}^{n-1}(S(u)\cap K)\,, \tag{5.68}$$

so that (5.66) and (5.67) yield

$$\limsup_{\varepsilon\to 0} F_\varepsilon(u_\varepsilon) \le a \int_\Omega |\nabla u(x)|^2\, dx + 2b\mathcal{H}^{n-1}(S(u)\cap K)\,.$$

As $K \searrow \overline{\Omega}$ we obtain

$$\limsup_{\varepsilon\to 0} F_\varepsilon(u_\varepsilon) \le a \int_\Omega |\nabla u(x)|^2\, dx + 2b\mathcal{H}^{n-1}(S(u)\cap \overline{\Omega})\,. \tag{5.69}$$

By (5.62) and (5.60) we have

$$\int_\Omega g_\varepsilon(x)\, dx \le \frac{1}{C_\Omega} \int_{R_\varepsilon} h_\varepsilon(x) dx + \frac{c^2}{\rho_\varepsilon^2} |M_\varepsilon|\, \|u\|^2_{L^\infty(\Omega')}\,. \tag{5.70}$$

Since $\big(|M_\varepsilon|/\varepsilon\big)$ is bounded by (5.68), from (5.59) and (5.70) we obtain

$$\lim_{\varepsilon\to 0} a_\varepsilon \int_\Omega \Big( ⨍_{B_\varepsilon(x)\cap\Omega} |\nabla u_\varepsilon(y)|^2\, dy \Big)\, dx = 0\,. \tag{5.71}$$

From (5.57), (5.69), and (5.71) it follows that

$$\mathcal{F}''(u) \leq \limsup_j F_{\varepsilon_h}(u_{\varepsilon_j}) \leq a\int_\Omega |\nabla u(x)|^2\,dx + 2b\mathcal{H}^{n-1}(S(u)\cap\overline{\Omega}),$$

which proves (5.58) and concludes the proof of the corollary. □

The previous result allows us to obtain the following corollary concerning the convergence of minimum points and of minimum values.

**Corollary 5.17** *Let $(f_\varepsilon)_{\varepsilon>0}$ be a family of non-decreasing continuous functions which satisfy conditions* (5.55) *and* (5.56) *of Corollary* 5.16 *for a suitable non-decreasing continuous function $f$ satisfying* (5.1). *Assume that for every $\varepsilon > 0$ we have*

$$\liminf_{t\to+\infty} \frac{f_\varepsilon(t)}{t} > 0\,. \tag{5.72}$$

*Then for every $g \in L^\infty(\Omega)$ and for every $\varepsilon > 0$ there exists a solution $u_\varepsilon$ to the minimum problem*

$$\min_{u\in H^1(\Omega)}\Big\{\frac{1}{\varepsilon}\int_\Omega f_\varepsilon\Big(\varepsilon \fint_{B_\varepsilon(x)\cap\Omega} |\nabla u(y)|^2\,dy\Big)\,dx + \int_\Omega |u(x)-g(x)|^p\,dx\Big\}, \tag{5.73}$$

*and for every sequence $(\varepsilon_j)$ of positive numbers converging to 0 as $j\to\infty$ there exists a subsequence (not relabelled) such that $(u_{\varepsilon_j})$ converges in $L^1(\Omega)$ to a solution $u_0$ of the problem*

$$\min_{u\in SBV(\Omega)}\Big\{a\int_\Omega |\nabla u(x)|^2\,dx + 2b\mathcal{H}^{n-1}(S(u)) + \int_\Omega |u(x)-g(x)|^p\,dx\Big\}. \tag{5.74}$$

*Furthermore the minimum value of problem* (5.73) *converges to the minimum value of problem* (5.74) *as $\varepsilon \to 0$.*

**Proof** Since $f$ is non-decreasing, inequality (5.72), together with (5.56) and (5.1), implies that for every $\varepsilon > 0$ there exists a constant $c_\varepsilon > 0$ such that

$$f_\varepsilon(t) \geq c_\varepsilon\, t \tag{5.75}$$

for every $t \geq 0$. Fix $\varepsilon > 0$ and let $(w_k)$ be a minimizing sequence for problem (5.73). By a truncation argument we may assume that $\|w_k\|_{L^\infty(\Omega)} \leq \|g\|_{L^\infty(\Omega)}$. From (5.75) and from Lemma 5.15 we deduce that the sequence $(w_k)$ is bounded in $H^1(\Omega)$. Passing to a subsequence, we may also assume that $(w_k)$ converges weakly in $H^1(\Omega)$ and strongly in $L^1(\Omega)$ to some function $w$. By the weak lower semicontinuity of the norm we obtain

$$\fint_{B_\varepsilon(x)\cap\Omega} |\nabla w(y)|^2\,dy \leq \liminf_{k\to\infty} \fint_{B_\varepsilon(x)\cap\Omega} |\nabla w_k(y)|^2\,dy$$

for every $x \in \Omega$, so that by Fatou's Lemma

$$\frac{1}{\varepsilon}\int_\Omega f_\varepsilon\Big(\varepsilon \fint_{B_\varepsilon(x)\cap\Omega} |\nabla w(y)|^2\,dy\Big)\,dx + \int_\Omega |w(x)-g(x)|^p\,dx \leq$$

$$\leq \liminf_{k\to\infty} \frac{1}{\varepsilon}\int_\Omega f_\varepsilon\Big(\varepsilon \fint_{B_\varepsilon(x)\cap\Omega} |\nabla w_k(y)|^2\,dy\Big)\,dx + \int_\Omega |w_k(x)-g(x)|^p\,dx\,.$$

This proves that $w$ is a minimum point of problem (5.73).

For every $\varepsilon > 0$ let $u_\varepsilon$ be a solution of problem (5.73). By a truncation argument it can be easily seen that $\|u_\varepsilon\|_{L^\infty(\Omega)} \leq \|g\|_{L^\infty(\Omega)}$. Moreover, by taking $u = 0$ in (5.73), we obtain the estimate

$$\frac{1}{\varepsilon}\int_\Omega f_\varepsilon\Big(\varepsilon \fint_{B_\varepsilon(x)\cap\Omega} |\nabla u_\varepsilon(y)|^2\,dy\Big)\,dx + \int_\Omega |u_\varepsilon(x)-g(x)|^p\,dx \leq \int_\Omega |g(x)|^p\,dx\,. \tag{5.76}$$

Let $(\varepsilon_j)$ be a sequence of positive numbers converging to 0 as $j\to\infty$. By Proposition 5.2 there exists sequence $(v_j)$ in $SBV(\Omega)\cap L^\infty(\Omega)$ such that

$$\int_\Omega |\nabla v_j|^2\,dx + \mathcal{H}^{n-1}(S(v_j)) + \|v_j\|_{L^\infty(\Omega)} \leq c\,, \tag{5.77}$$

$$\|v_j - u_{\varepsilon_j}\|_{L^1(\Omega)} \leq c\,\varepsilon_j + c\,|\Omega\setminus\Omega_{6\varepsilon_j}|\,, \tag{5.78}$$

for a suitable constant $c$ independent of $j$. By (5.77) and by Theorem 2.3 there exists a subsequence (not relabelled) such that $(v_j)$ converges in $L^1(\Omega)$ to some function $u_0 \in SBV(\Omega)$. By (5.78) the sequence $(u_{\varepsilon_j})$ converges to $u_0$ in $L^1(\Omega)$.

From the proof of Theorem 5.1 and of Corollary 5.16 it is clear that the $\Gamma$-convergence holds also with respect to the $L^p(\Omega)$ convergence. Let $G(u) = \int_\Omega |u-g|^p dx$; then the sequence $(\mathcal{F}_{\varepsilon_j} + G)$ $\Gamma$-converges to $\mathcal{F}+G$. Hence, the function $u_0$ is a minimum point of $\mathcal{F}+G$, i.e., a minimum point of problem (5.74), and $(\mathcal{F}_{\varepsilon_j}(u_{\varepsilon_j}) + G(u_{\varepsilon_j}))$ converges to $\mathcal{F}(u_0)+G(u_0)$. This proves of the convergence of the minimum values. □

**Remark 5.18** Theorem 5.1 and Corollaries 5.16 and 5.17 still hold, with obvious modifications in the proofs, if the term

$$\varepsilon \fint_{B_\varepsilon(x)\cap\Omega} |\nabla u(y)|^2\,dy$$

is replaced by

$$\frac{\varepsilon}{|B_\varepsilon|}\int_{B_\varepsilon(x)\cap\Omega} |\nabla u(y)|^2\,dy \tag{5.79}$$

in the definitions of $F_\varepsilon$ and $\mathcal{F}_\varepsilon$, and in problem (5.73). In the proof of the existence of a solution of (5.73) one can use the inequality containing (5.79) in Lemma 5.15.

## 5.2 Finite-difference approximation of the Mumford-Shah functional

In this section we prove a $n$-dimensional finite-difference approximation of the Mumford-Shah functional, which generalizes the 1-dimensional result of Section 3.5.

Let $\rho : \mathbf{R}^n \to [0, +\infty)$ be a symmetric mollifier (i.e., $\rho(x) = \psi(|x|)$), and let $f : [0, +\infty) \to [0, +\infty)$ be a Borel function with $f(0) = 0$, such that for all $c > 0$ $\inf\{f(t) : t \geq c\} > 0$, $a, b > 0$ exist with

$$\lim_{t \to 0+} \frac{f(t)}{t} = a, \qquad \lim_{t \to +\infty} f(t) = b,$$

and $f(t) \leq \min\{at, b\}$.

We define the functionals $F_\varepsilon : L^1_{\mathrm{loc}}(\mathbf{R}^n) \to [0, +\infty)$ as

$$F_\varepsilon(u) = \frac{1}{\varepsilon} \int_{\mathbf{R}^n} \int_{\mathbf{R}^n} f\Big(\frac{(u(y) - u(x))^2}{\varepsilon}\Big) \rho_\varepsilon(y - x)\, dy\, dx\,. \tag{5.80}$$

The introduction of the convolution kernel $\rho$ has been proposed by De Giorgi to overcome the anisotropy that obviously results if we take difference quotients only in the coordinate directions (see e.g. [Ch], [BGe]).

A simple change of variables yields

$$F_\varepsilon(u) = \frac{1}{\varepsilon} \int_{\mathbf{R}^n} \int_{\mathbf{R}^n} f\Big(\frac{(u(x + \varepsilon\xi) - u(x))^2}{\varepsilon}\Big) \rho(\xi)\, dx\, d\xi\,. \tag{5.81}$$

**Theorem 5.19** *The functionals $F_\varepsilon$ $\Gamma$-converge as $\varepsilon \to 0^+$ with respect to the $L^1_{\mathrm{loc}}(\mathbf{R}^n)$-convergence to the Mumford-Shah functional $F$ defined by*

$$F(u) = \begin{cases} A \displaystyle\int_{\mathbf{R}^n} |\nabla u|^2\, dx + B\mathcal{H}^{n-1}(S(u)) & \text{if } u \in GSBV_{\mathrm{loc}}(\mathbf{R}^n) \\ +\infty & \text{otherwise,} \end{cases} \tag{5.82}$$

*where $A, B$ are defined by*

$$A = a\omega_n \int_0^{+\infty} t^{n+1} \psi(t)\, dt, \qquad B = 2b\omega_{n-1} \int_0^{+\infty} t^n \psi(t)\, dt \tag{5.83}$$

*(where $\rho(\xi) = \psi(|\xi|)$).*

**Proof** The liminf inequality will be obtained by a slicing procedure which allows to reduce to the 1-dimensional case.

Using the notation of Section 1.8.1 we remark that, if we set

$$F^1_\varepsilon(v) = \frac{1}{\varepsilon} \int_{\mathbf{R}} f\Big(\frac{(v(t + \varepsilon) - u(t))^2}{\varepsilon}\Big)\, dt,$$

then

$$F_\varepsilon(u) = \int_{\mathbf{R}^n} \int_{\Pi_\xi} |\xi| F^1_\varepsilon(u_{\xi,y}) d\mathcal{H}^{n-1}(y) \rho(\xi)\, d\xi\,. \tag{5.84}$$

Let $u_j \to u$ in $L^1_{\mathrm{loc}}(\mathbf{R}^n)$ and let $\varepsilon_j \to 0^+$. By Fatou's Lemma, we then get

$$\liminf_j F_{\varepsilon_j}(u_j) \geq \int_{\mathbf{R}^n} \int_{\Pi_\xi} |\xi| \liminf_j F^1_{\varepsilon_j}(u_{j\,\xi,y}) d\mathcal{H}^{n-1}(y) \rho(\xi)\, d\xi\,.$$

By Theorem 3.41 we know that the sequence $(F^1_{\varepsilon_j})$ $\Gamma$-converges to the Mumford-Shah functional $F^1$ whose value on $SBV(\mathbf{R})$ is

$$F^1(v) = a \int_{\mathbf{R}} |v'|^2\, dt + b\#(S(v))\,.$$

We deduce that if $\liminf_j F_{\varepsilon_j}(u_j) < +\infty$ then for all $\xi \in \mathbf{R}^n$ and for $\mathcal{H}^{n-1}$-a.a. $y \in \Pi_\xi$ we have $u_{\xi,y} \in SBV(\mathbf{R})$, and

$$\int_{\Pi_\xi} \int_{\mathbf{R}} |u'_{\xi,y}|^2\, dt + b\#(S(u_{\xi,y})) d\mathcal{H}^{n-1}(y) < +\infty\,.$$

By Theorem 4.1(b) we deduce that $u \in GSBV_{\rm loc}(\mathbf{R}^n)$. Again from Theorem 4.1(a) we then have

$$\begin{aligned}\liminf_j F_{\varepsilon_j}(u_j) &\geq \int_{\mathbf{R}^n} \int_{\Pi_\xi} |\xi| F^1(u_{\xi,y}) d\mathcal{H}^{n-1}(y) \rho(\xi)\, d\xi \\ &= \int_{\mathbf{R}^n} \Big( \int_{\mathbf{R}^n} a |\langle \nabla u, \xi\rangle|^2\, dx + b \int_{S(u)} |\langle \nu_u, \xi\rangle| d\mathcal{H}^{n-1}\Big) \rho(\xi)\, d\xi\,.\end{aligned}$$

Now, if we remark that

$$\int_{\mathbf{R}^n} |\langle \eta, \xi\rangle|^2 \rho(\xi)\, d\xi = |\eta|^2 \frac{1}{n} \int_{\mathbf{R}^n} |\xi|^2 \rho(\xi)\, d\xi$$

for all $\eta \in \mathbf{R}^n$, we get

$$\begin{aligned}\int_{\mathbf{R}^n} \int_{\mathbf{R}^n} a |\langle \nabla u, \xi\rangle|^2\, dx \rho(\xi)\, d\xi &= \int_{\mathbf{R}^n} \int_{\mathbf{R}^n} a |\langle \nabla u, \xi\rangle|^2 \rho(\xi)\, d\xi\, dx \\ &= a \frac{1}{n} \int_{\mathbf{R}^n} |\xi|^2 \rho(\xi)\, d\xi \int_{\mathbf{R}^n} |\nabla u|^2\, dx\,,\end{aligned}$$

and the expression for $A$ follows after a simple computation.

As for the term $B\mathcal{H}^{n-1}(S(u))$ in the definition of $F$, we have

$$\begin{aligned}&\int_{\mathbf{R}^n} \int_{S(u)} |\langle \nu_u, \xi\rangle| d\mathcal{H}^{n-1} |\xi| \rho(\xi)\, d\xi \\ &= \int_0^{+\infty} t^n \psi(t) \int_{S^{n-1}} \int_{S(u)} |\langle \nu_u, \xi\rangle| d\mathcal{H}^{n-1}\, d\xi\, dt \\ &= \int_0^{+\infty} t^n \psi(t)\, dt \int_{S(u)} \int_{S^{n-1}} |\langle \nu_u, \xi\rangle| d\mathcal{H}^{n-1}\, d\xi,\end{aligned}$$

and we recover the value of $B$ by remarking that $\int_{S^{n-1}} |\langle \nu, \xi\rangle|\,d\xi = 2\omega_{n-1}$ for all $\nu \in S^{n-1}$.

It is sufficient to exhibit a recovery sequence for $u \in GSBV_{\rm loc}(\mathbf{R}^n)$. In this case we trivially take $u_\varepsilon = u$ for all $\varepsilon > 0$, and use Remark 3.42 together with (5.84), to obtain

$$F_\varepsilon(u) \le \int_{\mathbf{R}^n} \int_{\Pi_\xi} |\xi| F^1(u_{\xi,y}) d\mathcal{H}^{n-1}(y) \rho(\xi)\, d\xi\,.$$

Eventually we proceed as above to see that the right-hand side equals $F(u)$, and obtain that $F_\varepsilon(u) \le F(u)$ for all $\varepsilon > 0$. □

### 5.2.1 *Compactness*

We prove a compactness result for sequences of functions $(u_j)$ with $\|u_j\|_\infty + F_{\varepsilon_j}(u_j)$ uniformly bounded. Note that this result does not hold for the 1-dimensional functionals of Section 3.5.

Since there exists a constant $c > 0$ such that $f(t) \ge c\min\{t, 1\}$, it will be sufficient to treat the case

$$f(t) = \min\{t, 1\}\,.$$

Let

$$F_{\varepsilon,\xi}(u) = \frac{1}{\varepsilon}\int_{\mathbf{R}^n} f\Big(\frac{(u(x+\varepsilon\xi) - u(x))^2}{\varepsilon}\Big)\, dx$$

be defined for $\varepsilon > 0$, $\xi \in \mathbf{R}^n$ and $u \in L^1_{\rm loc}(\mathbf{R}^n)$.

**Lemma 5.20** *Let $u \in L^\infty(\mathbf{R}^n)$; then*

$$\int_A |u(x+\delta\xi) - u(x)|\, dx \le 3\delta(|A|^{1/2} + \|u\|_\infty)(1 + F_{\delta,\xi}(u)) \tag{5.85}$$

*for all $\delta > 0$, $\xi \in \mathbf{R}^n$ and $A \subset\subset \mathbf{R}^n$.*

**Proof** Let $A_\delta = \{x \in A:\ |u(x+\delta\xi) - u(x)|^2 > \delta\}$. Using Hölder's inequality, and the fact that $f(z) = z$ if $|z| \le 1$, we get

$$\begin{aligned}\int_{A\setminus A_\delta} |u(x+\delta\xi) - u(x)|\, dx &\le \delta|A|^{1/2}\Big(\frac{1}{\delta}\int_{A\setminus A_\delta} f\Big(\frac{|u(x+\delta\xi) - u(x)|^2}{\delta}\Big)\, dx\Big)^{1/2}\\ &\le \delta|A|^{1/2}(F_{\delta,\xi}(u))^{1/2} \le \delta|A|^{1/2}(1 + F_{\delta,\xi}(u))\,.\end{aligned}$$

Since trivially

$$|A_\delta| = \int_{A_\delta} f\Big(\frac{|u(x+\delta\xi) - u(x)|^2}{\delta}\Big)\, dx \le \delta F_{\delta,\xi}(u),$$

so that

$$\int_{A_\delta} |u(x+\delta\xi)-u(x)|\,dx \le 2\|u\|_\infty |A_\delta| \le 2\delta\|u\|_\infty F_{\delta,\xi}(u),$$

the thesis is easily proven. □

**Lemma 5.21** *Let $u \in L^\infty(\mathbf{R}^n)$; then*

$$\int_A |u*\rho_\delta - u|\,dx \le 3\delta(|A|^{1/2}+\|u\|_\infty)(1+F_\delta(u))\,. \tag{5.86}$$

*for all $\delta>0$, $\xi\in\mathbf{R}^n$ and $A \subset\subset \mathbf{R}^n$.*

**Proof** As we can write

$$F_\delta(u) = \int_{\mathbf{R}^n} F_{\delta,\xi}(u)\rho(\xi)\,d\xi,$$

by the previous lemma we have

$$\begin{aligned}\int_A |u*\rho_\delta(x)-u(x)|\,dx &\le \int_{A\times\mathbf{R}^n} |u(x+\delta\xi)-u(x)|\rho(\xi)\,d\xi\,dx\\ &\le 3\delta(|A|^{1/2}+\|u\|_\infty)\int_{\mathbf{R}^n}(1+F_{\delta,\xi}(u))\rho(\xi)\,d\xi\\ &= 3\delta(|A|^{1/2}+\|u\|_\infty)(1+F_\delta(u))\,,\end{aligned}$$

as required. □

**Lemma 5.22** *Let $u \in L^\infty(\mathbf{R}^n)$; then*

$$F_{k\varepsilon}(u) \le F_\varepsilon(u) \tag{5.87}$$

*for all $\varepsilon>0$ and for all $k \in \mathbf{N}\setminus\{0\}$.*

**Proof** We argue by induction. The thesis is trivial if $k=1$. Suppose it holds for $k\ge 1$. Since we can write

$$\begin{aligned}&\frac{(u(x+(k+1)\varepsilon\xi)-u(x))^2}{(k+1)\varepsilon}\\ &= \frac{(u(x+(k+1)\varepsilon\xi)-u(x+k\varepsilon\xi)+u(x+k\varepsilon\xi)-u(x))^2}{(k+1)\varepsilon}\\ &\le \frac{(u(x+(k+1)\varepsilon\xi)-u(x+k\varepsilon\xi))^2}{\varepsilon}+\frac{(u(x+k\varepsilon\xi)-u(x))^2}{k\varepsilon},\end{aligned}$$

and $f$ is increasing and subadditive, we have

$$\frac{1}{(k+1)\varepsilon} f\Big(\frac{(u(x+(k+1)\varepsilon\xi)-u(x))^2}{(k+1)\varepsilon}\Big)$$

$$\leq \frac{1}{(k+1)\varepsilon} f\Big(\frac{(u(x+(k+1)\varepsilon\xi) - u(x+k\varepsilon\xi))^2}{\varepsilon}\Big) + \frac{1}{(k+1)\varepsilon} f\Big(\frac{(u(x+k\varepsilon\xi) - u(x))^2}{k\varepsilon}\Big).$$

Multiplying by $\rho(\xi)$ and integrating, we get

$$F_{(k+1)\varepsilon}(u) \leq \frac{1}{k+1} F_\varepsilon(u) + \frac{k}{(k+1)} F_{k\varepsilon}(u) \leq F_\varepsilon(u),$$

as the last inequality follows by the inductive hypothesis. □

**Theorem 5.23** *Let $\varepsilon_j \to 0$ and let $u_j \in L^\infty(\mathbf{R}^n)$ be such that*

$$\sup_j \Big(F_{\varepsilon_j}(u_j) + \|u_j\|_\infty\Big) < +\infty. \tag{5.88}$$

*Then there exists a subsequence (not relabelled) of $(u_j)$ and $u \in GSBV_{\rm loc}(\mathbf{R}^n)$ such that $u_j \to u$ in $L^1_{\rm loc}(\mathbf{R}^n)$.*

**Proof** It will suffice to show that for every $A \subset\subset \mathbf{R}^n$ there exists a constant $M$ such that for every $\sigma > 0$ there exists a sequence $(v_j)$ pre-compact in $L^1(A)$ with

$$\|u_j - v_j\|_{L^1(A)} \leq M\sigma.$$

Let $k_j = [\sigma/\varepsilon_j]$ and let

$$v_j = \begin{cases} u_j * \rho_{k_j\varepsilon_j} & \text{if } \varepsilon_j < \sigma/2 \\ u_j & \text{otherwise.} \end{cases}$$

To see that $(v_j)$ is pre-compact it suffices to consider indices $j$ with $\varepsilon_j < \sigma/2$. Note that $k_j\varepsilon_j \geq \sigma/2$ so that the $C^1$ norms of $v_j$ are equibounded on $\mathbf{R}^n$. By Ascoli's Theorem $(v_j)$ is pre-compact in $C^0(A)$, and so also in $L^1(A)$.

By Lemmas 5.21 and 5.22 we have

$$\begin{aligned}\|u_j - v_j\|_{L^1(A)} &\leq 3\sigma(|A|^{1/2} + \|u_j\|_\infty)(1 + F_{k_j\varepsilon_j}(u_j)) \\ &\leq 3\sigma(|A|^{1/2} + \|u_j\|_\infty)(1 + F_{\varepsilon_j}(u_j)),\end{aligned}$$

and we can take $M = 3\sup_j(|A|^{1/2} + \|u_j\|_\infty)(1 + F_{\varepsilon_j}(u_j))$ to obtain the thesis. Note that $F(u) < +\infty$, so that $u \in GSBV_{\rm loc}(\mathbf{R}^n)$ □

### 5.2.2 *Convergence results*

As the functionals $F_\varepsilon$ are not coercive, we have to introduce an additional constraint to obtain the existence of approximate solutions to problems involving the Mumford-Shah functional.

**Theorem 5.24** *Let $1 \le p < +\infty$, and let $g \in L^p(\mathbf{R}^n) \cap L^\infty(\mathbf{R}^n)$. Then for every $\varepsilon > 0$ there exists a solution $u_\varepsilon$ to the minimum problem*

$$m_\varepsilon = \min\Big\{F_\varepsilon(u) + \int_{\mathbf{R}^n} |u-g|^p\,dx : u \in BV(\mathbf{R}^n), |Du| \le 1/\varepsilon\Big\},$$

*and for every sequence $\varepsilon_j \to 0^+$ there exists a subsequence (not relabelled) converging to a solution $u$ of the problem*

$$m_0 = \min_{u \in SBV(\mathbf{R}^n)}\Big\{A\int_{\mathbf{R}^n} |\nabla u|^2\,dx + B\mathcal{H}^{n-1}(S(u)) + \int_{\mathbf{R}^n} |u-g|^p\,dx\Big\}.$$

*Furthermore, $m_\varepsilon \to m_0$ as $\varepsilon \to 0^+$*

**Proof** See Exercises 5.2–5.5 below. □

5.2.3 *Exercises*

**Exercise 5.1** Compute the Γ-limit in Theorem 5.19 when $f(z) = \arctan z$ and $\rho(y) = c\exp(-y^2)$.

**Exercise 5.2** Prove that the Γ-convergence of $F_\varepsilon$ to $F$ holds also with respect to the $L^p_{\rm loc}(\mathbf{R}^n)$ convergence, provided the functionals are defined in $L^p_{\rm loc}(\mathbf{R}^n)$ ($1 \le p < +\infty$) (repeat the reasoning of Theorem 3.33).

**Exercise 5.3** Prove that the functional $F_\varepsilon$ is lower semicontinuous in $L^1_{\rm loc}(\mathbf{R}^n)$ (use Fatou's Lemma).

**Exercise 5.4** Let $1 \le p < +\infty$, and let $g \in L^p(\mathbf{R}^n) \cap L^\infty(\mathbf{R}^n)$. Show that the functionals $G_\varepsilon : L^1_{\rm loc}(\mathbf{R}^n) \to [0, +\infty]$

$$G_\varepsilon(u) = \begin{cases} F_\varepsilon(u) + \displaystyle\int_{\mathbf{R}^n} |u-g|^p\,dx & \text{if } u \in BV(\mathbf{R}^n),\ |Du|(\mathbf{R}^n) \le 1/\varepsilon \\ +\infty & \text{otherwise} \end{cases}$$

Γ-converge to $F(u) + \int_{\mathbf{R}^n} |u-g|^p\,dx$.

*Hint*: The liminf inequality is trivial. Prove the pointwise convergence (and, hence, also exhibit a recovery sequence) if $u \in SBV(\mathbf{R}^n)$. For a general $u$ reason by truncation. Note that the Γ-limit is finite only if $u \in L^p(\mathbf{R}^n)$.

**Exercise 5.5** Prove Theorem 5.24.

*Hint*: By truncation restrict to the set $\{\|v\|_\infty \le \|g\|_\infty\}$. Use the compactness of $\{v \in BV_{\rm loc}(\mathbf{R}^n) : \|v\|_\infty \le \|g\|_\infty, |Dv| \le 1/\varepsilon\}$ with respect to the $L^1_{\rm loc}(\mathbf{R}^n)$-convergence to obtain existence of minimizers $u_\varepsilon$. Use Theorem 5.23 to obtain compactness and the properties of Γ-convergence to obtain convergence of minima and minimizers.

# APPENDIX A

## SOME NUMERICAL RESULTS

The methods studied in these lecture notes provide approximations of free-discontinuity problems through functionals defined on smooth functions. These functionals themselves can be approximated through a discretization procedure. The problems related to this kind of issues are not relevant to our presentation, and will not be discussed in detail in this book. However, we include for the sake of illustration two numerical results.

First, we recall that a *j-dimensional simplex* in $\mathbf{R}^n$ ($j \in \{1, 2, \ldots, n\}$) is the convex hull of $j+1$ points $x_0, x_1, \ldots, x_j$ (called the *vertices* of the simplex) which are not contained in a hyperplane of dimension $j-1$. The *faces* of a $j$-dimensional simplex are the $(j-1)$-dimensional simplexes generated by any $j$ of its vertices. For every simplex $K \subseteq \mathbf{R}^n$, we denote $\delta_K$ the diameter of $K$ and $\varrho_K$ the *inner radius* of $K$, that is, the supremum of the diameters of the $j$-dimensional balls contained in $K$.

For the rest of this appendix, we will assume that the set $\Omega$ is a polyhedron. We say that a finite family $T = \{K\}_{K \in T}$ of $n$-dimensional simplexes is a *triangulation* of $\Omega$ if the following conditions are satisfied:

(T1) $\overline{\Omega} = \bigcup_{K \in T} K$;

(T2) if $K_1, K_2 \in T$ and $K_1 \neq K_2$, then $\mathring{K}_1 \cap \mathring{K}_2 = \emptyset$;

(T3) for every $K \in T$, any face of $K$ is either contained in $\partial\Omega$, or it is also a face of a simplex of $T$ different from $K$.

The vertices of the simplexes $K \in T$ are called the *nodes* of $T$.

Let $T$ be a triangulation of $\Omega$; we call $PA(T)$ the vector space of all the continuous functions on $\Omega$ whose restriction to $K$ is an affine function for every $K \in T$. It can be easily seen that, if $\phi$ is any real-valued function defined on the nodes of $T$, then there exists one and only one function in $PA(T)$ which takes on each node the value prescribed by $\phi$.

From now on, $\mathcal{T} = \{T_h\}_{h>0}$ will be a fixed family of triangulations of $\Omega$. We will assume that $\mathcal{T}$ is *regular*; i.e., there exist two positive constants $c_1, c_2$ such that

(R1) $\delta_K \le c_1 h$ for every $K \in T_h$ and for every $h > 0$;

(R2) $\delta_K \le c_2 \varrho_K$ for every $K \in T_h$ and for every $h > 0$.

For each $h$ we consider the *interpolation operator* associated to $T_h$, $\Pi_h : C^0(\overline{\Omega}) \longrightarrow PA(T_h)$, which associates to $u$ the piecewise affine function which agrees with $u$ at the nodes of $T_h$. The function $\Pi_h u$ is called the *$T_h$-interpolant* of $u$.

Let $g \in L^\infty(\Omega)$ be fixed, and let $C > 0$. We shall be concerned with the approximation of the Mumford-Shah functional with a lower order term

$$F(u) = \begin{cases} \displaystyle\int_\Omega |\nabla u|^2\,dx + \mathcal{H}^{n-1}(S(u)) + C\int_\Omega |u-g|^2\,dx & \text{if } u \in GSBV(\Omega) \\ +\infty & \text{otherwise,} \end{cases}$$

defined on $L^1(\Omega)$.

**Elliptic discrete approximation**

As a first example, we illustrate a discrete version of the approximation of the Mumford-Shah by elliptic functionals. For a proof we refer to [BeC].

For all $\varepsilon > 0$, let $g_\varepsilon \in C_0^\infty(\Omega)$ be such that

$$g_\varepsilon \to g \text{ in } L^2(\Omega), \qquad \|g_\varepsilon\|_\infty \le \|g\|_\infty, \qquad \|\nabla g_\varepsilon\|_\infty \le \frac{c}{\varepsilon}.$$

Let $W(s) = 1 - s^2$ and let $k(\varepsilon) = o(\varepsilon)$. We define the functionals

$$G_{\varepsilon,h}(u,v) = \begin{cases} \displaystyle\int_\Omega \Big((v + k(\varepsilon))|\nabla u|^2 + \frac{2}{\pi}\Big(\varepsilon|\nabla v|^2 + \frac{1}{4\varepsilon}\Pi_h(W(v))\Big)\Big)\,dx & \\ \qquad\qquad \displaystyle +C\int_\Omega \Pi_h(|u-g_\varepsilon|^2) \quad \text{if } u, v \in PA(T_h) \text{ and } 0 \le v \le 1, & \\ +\infty \qquad\qquad\qquad\qquad \text{otherwise.} & \end{cases}$$

**Theorem A.1** *If $h(\varepsilon) = o(\varepsilon)$ then we have*

$$\Gamma(L^2(\Omega))\text{-}\lim_{\varepsilon\to 0+} G_{\varepsilon,h(\varepsilon)}(u,1) = F(u)$$

*for all $u \in SBV(\Omega) \cap L^\infty(\Omega)$. Moreover, if $(\varepsilon_j)$ is a sequence of positive numbers converging to 0, and $(u_j, v_j)$ is a family of absolute minimizers of $G_{\varepsilon_j,h(\varepsilon_j)}$ then $v_j \to 1$, $(u_j)$ is relatively compact in $L^2(\Omega)$, and every limit of a subsequence of $(u_j)$ is a minimizer of $F$.*

**Non-local discrete approximation**

We now illustrate a discrete version of the non-local approximation of the Mumford-Shah functional. For a proof, we refer to [Co1].

First, we introduce a more detailed notation for the triangulations $T_h$. For every fixed $h > 0$, we index the finite elements that belong to $T_h$ by a parameter varying in a finite set $I_h$, so that we can write

$$T_h = \{K_i^h : i \in I_h\}.$$

For every $\varepsilon > 0$, $h > 0$ and $i \in I_h$, we define the set of indices

$$J^i_{\varepsilon,h} = \{j \in I_h : K^h_j \cap (K^h_i + B_\varepsilon(0)) \neq \emptyset\},$$

where $K^h_i + B_\varepsilon(0)$ is simply the $\varepsilon$-neighborhood of $K^h_i$; we also set

$$P^i_{\varepsilon,h} = \Omega \cap \bigcup_{j \in J^i_{\varepsilon,h}} K^h_j .$$

For almost every $x \in \Omega$, there exists a unique $i(x) \in I_h$ such that $x \in K^h_{i(x)}$. Define $P_{\varepsilon,h}(x) := P^{i(x)}_{\varepsilon,h}$, and

$$F_{\varepsilon,h}(u) = \begin{cases} \dfrac{1}{\varepsilon}\displaystyle\int_\Omega f\Big(\frac{\varepsilon}{|B_\varepsilon(0)|}\int_{P_{\varepsilon,h}(x)} |Du(y)|^2\,dy\Big)\,dx + C\int_\Omega |u-g_h|^2\,dx & \\ & \text{if } u \in PA(T_h) \\ +\infty & \text{if } u \in L^2(\Omega)\setminus PA(T_h), \end{cases}$$

where for every $h > 0$, the function $g_h$ is defined as

$$g_h(x) = \sum_{K \in T_h} \Big(\frac{1}{|K|}\int_K g(y)\,dy\Big)\,\chi_K(x).$$

**Theorem A.2** *Let $h(\varepsilon) = o(\varepsilon)$; then the family $(F_{\varepsilon,h(\varepsilon)})$ $\Gamma$-converges to $F$ in the strong topology of $L^2(\Omega)$ as $\varepsilon \to 0$. Moreover, if $u_\varepsilon$ denotes a solution of the minimum problem*

$$m_\varepsilon = \min\{F_{\varepsilon,h(\varepsilon)}(u) : u \in PA(T_{h(\varepsilon)})\},$$

*and $(\varepsilon_j)$ is any sequence of positive numbers converging to 0, then the sequence $(u_{\varepsilon_j})$ has a subsequence which converges strongly in $L^2(\Omega)$ to a solution of*

$$m = \min\{F(u) : u \in SBV(\Omega)\},$$

*and $m = \lim_{\varepsilon\to 0^+} m_\varepsilon$.*

# APPENDIX B

## APPROXIMATION OF POLYHEDRAL ENERGIES

In this appendix we outline another application of the approximation method described in Section 3.2.1 to the problem of variational approximation of energies defined on polyhedra with fixed possible orientations. We will restrict our analysis to a symmetric case, where these polyhedra are simply coordinate polyrectangles. For the non-symmetric case we refer to [BM].

For every $\varepsilon > 0$ and for every $E \subseteq \mathbf{R}^2$ of class $C^2$, we define the energy

$$F_\varepsilon(E) = \int_{\partial E} \Big(\frac{1}{\varepsilon}\varphi(\nu) + \varepsilon\kappa^2\Big) d\mathcal{H}^1, \tag{B.1}$$

where $\nu = \nu(x)$ is the outer unit normal to $\partial E$, $\kappa = \kappa(x)$ is the curvature of $\partial E$ at $x \in \partial E$, and $\varphi : S^1 \to [0, +\infty)$ is of class $C^2$, symmetric with respect to the axes and the bisectrices, and such that

$$\varphi(\nu) = 0 \Longleftrightarrow \nu = \pm e_i, \qquad i = 1, 2. \tag{B.2}$$

We set $W(t) = \varphi(t, \sqrt{1-t^2})$ for $t \in [0,1]$, and we suppose in addition that $W''(0) > 0$. We define

$$c_W = 2\int_0^1 \sqrt{\frac{W(\tau)}{1-\tau^2}} d\tau. \tag{B.3}$$

**Theorem B.1** *Let $G_\varepsilon : \mathrm{L}^1(\mathbf{R}^2) \to [0, +\infty]$ be defined by*

$$G_\varepsilon(u) = \begin{cases} F_\varepsilon(E) & \text{if } u = \chi_E \text{ with } E \text{ of class } C^2 \\ +\infty & \text{otherwise.} \end{cases} \tag{B.4}$$

*Then $G_\varepsilon$ $\Gamma$-converges as $\varepsilon \to 0$, with respect to the $\mathrm{L}^1_{\mathrm{loc}}(\mathbf{R}^2)$-convergence to the functional*

$$G(u) = \begin{cases} F(E) & \text{if } u = \chi_E \text{ with } E \text{ finite union of coordinate rectangles} \\ +\infty & \text{otherwise,} \end{cases} \tag{B.5}$$

*where we have set $F(E) = c_W \#(\text{corners of } E)$ for each $E$ finite union of coordinate rectangles.*

The proof of Theorem B.1 will be obtained at the end of the section, as a consequence of some previous propositions. In the following, $Q(x_0, R)$ denotes the cube centred in $x_0$, with sides of length $R$ and parallel to the axes. If $E_j, E$ are sets, we write $E_j \to E$ meaning that $\chi_{E_j} \to \chi_E$ in $\mathrm{L}^1_{\mathrm{loc}}(\mathbf{R}^2)$.

**Proposition B.2** *Let $a, b, \delta \in \mathbf{R}$, with $a < b$, and $0 < \delta < \frac{1}{2}$. Then for every $\eta : [a,b] \to S^1$ of class $C^1$ such that $\eta(a) = (\delta, \sqrt{1-\delta^2})$ and $\eta(b) = (\sqrt{1-\delta^2}, \delta)$ we have*

$$\int_a^b \Big(\frac{1}{\varepsilon}\varphi(\eta(t)) + \varepsilon|\dot\eta(t)|^2\Big)dt \geq c_W + O(\sqrt{\delta}). \tag{B.6}$$

**Proof** Setting $u(t) = \eta_1(t)$, we get

$$\int_a^b \Big(\frac{1}{\varepsilon}\varphi(\eta(t)) + \varepsilon|\dot\eta(t)|^2\Big)dt = \int_a^b \Big(\frac{1}{\varepsilon}W(u(t)) + \varepsilon\frac{(u'(t))^2}{1-u(t)^2}\Big)dt$$

$$\geq 2\int_a^b \sqrt{\frac{W(u(t))}{1-u(t)^2}}|u'(t)|dt \geq \int_\delta^{\sqrt{1-\delta^2}} \sqrt{\frac{W(\tau)}{1-\tau^2}}d\tau \geq \int_0^1 \sqrt{\frac{W(\tau)}{1-\tau^2}}d\tau + O(\sqrt{\delta}),$$

which is the desired inequality. □

**Remark B.3** Let $a, b, \delta$ be as in Proposition B.2. Then for every $\eta : [a,b] \to S^1$ of class $C^1$ such that $\delta \leq \eta_1(t) \leq \sqrt{1-\delta^2}$ for all $t \in [a,b]$ we have

$$\int_a^b \Big(\frac{1}{\varepsilon}\varphi(\eta(t)) + \varepsilon|\dot\eta(t)|^2\Big)dt \geq \frac{1}{\varepsilon}(b-a) \sup_{\tau\in[\delta,\sqrt{1-\delta^2}]} W(\tau). \tag{B.7}$$

Now we fix a sequence $(\varepsilon_j)$ of positive numbers converging to 0, and $E_j$ a sequence of sets with $\sup_j F_{\varepsilon_j}(E_j) < +\infty$. Suppose for the beginning that for every $j$, $\partial E_j$ has just one connected component. If we parameterize it by arc-length by a curve $\gamma_j : [0, T_j] \to \mathbf{R}^2$, we get

$$F_{\varepsilon_j}(E_j) = \int_0^{T_j} \Big(\frac{1}{\varepsilon_j}\varphi(\dot\gamma_j(t)) + \varepsilon_j|\ddot\gamma_j(t)|^2\Big)dt. \tag{B.8}$$

Fixed $0 < \delta < \frac{1}{2}$, we set

$$I_\delta^j = \Big\{t \in [0,T_j] : \delta < |\dot\gamma_1(t)| < \sqrt{1-\delta^2}\Big\}; \qquad J_\delta^j = [0,T_j] \setminus I_\delta^j.$$

We can suppose that $0 \in J_\delta^j$, so that $I_\delta^j$ is the union of an at most countable family of open intervals, which are the components of $I_\delta^j$. We denote by $I_\delta^{k,j} = (a_\delta^{k,j}, b_\delta^{k,j})$, $k \in K_\delta^j$, those components of $I_\delta^j$ for which $\dot\gamma_j(a_\delta^{k,j}) \neq \dot\gamma_j(b_\delta^{k,j})$. We can suppose moreover that $0 \in I_\delta^{k,j}$ for some $k$.

**Proposition B.4** *For every $\rho > 0$ and every $R > 0$, there exist $\delta > 0$ and $j_0 \in \mathbf{N}$ with the following property: let $k \in K_\delta^j$, and let $(r,s)$ be a component of $[0,T_j] \setminus \bigcup_{j\in K_\delta^j\setminus k} I_\delta^{k,j}$. Then for every $t \in (r,s)$ such that $\gamma_j(t) \in Q(\gamma(r), R)$, we have*

$$|\gamma_{j,2}(t) - \gamma_{j,2}(r)| < \rho \quad \text{if} \quad |\dot\gamma_{j,1}(r)| = \sqrt{1-\delta^2};$$
$$|\gamma_{j,1}(t) - \gamma_{j,1}(r)| < \rho \quad \text{if} \quad |\dot\gamma_{j,1}(r)| = \delta.$$

**Proof** Clearly, it suffices to consider the case $R > \rho$. We prove the proposition only for $\dot{\gamma}_{j,1}(t) = \sqrt{1-\delta^2}$, the other cases being completely analogous. From (B.7) it follows that $|I^j_\delta| \to 0$, as $j \to +\infty$, so we can choose $\delta$ and $j_0$ such that

$$\frac{\delta(R+\frac{\rho}{2})}{\sqrt{1-\delta^2}} < \rho, \qquad |I^j_\delta| < \frac{\rho}{2} \tag{B.9}$$

for all $j \geq j_0$. Note that by the construction of $I^j_\delta$, we must have

$$\dot{\gamma}_{j,1}(t) > \delta \text{ for all } t \in (r,s), \quad \text{and} \quad \dot{\gamma}_{j,1}(t) > \sqrt{1-\delta^2} \text{ for all } t \in (r,s) \setminus I^j_\delta,$$

so that

$$\begin{aligned}\gamma_{j,1}(t) = \gamma_{j,1}(r) + \int_r^t \dot{\gamma}_{j,1}(y)dy &\geq \gamma_{j,1}(r) + \int_{(r,t)\setminus I^j_\delta} \dot{\gamma}_{j,1}(y)dy \\ &\geq \gamma_{j,1}(r) + (t - r - \frac{\rho}{2})\sqrt{1-\delta^2},\end{aligned}$$

for $t \in (r,s)$. In order to have $\gamma_j(t)$ in $Q(\gamma_j(r), R)$, it is necessary that

$$t - r \leq \frac{R}{2\sqrt{1-\delta^2}} + \frac{\rho}{2} \leq \frac{R+\frac{\rho}{2}}{2\sqrt{1-\delta^2}}.$$

This inequality and our choices in (B.9) imply that

$$\begin{aligned}|\gamma_{j,2}(t) - \gamma_{j,2}(r)| &\leq \int_r^t |\dot{\gamma}_{j,2}(y)|dy \\ &\leq \int_{(r,t)\setminus I^j_\delta} \dot{\gamma}_{j,2}(y)dy + \frac{\rho}{2} \leq \frac{\delta(R+\frac{\rho}{2})}{2\sqrt{1-\delta^2}} + \frac{\rho}{2} < \rho,\end{aligned}$$

for all $t \in \left(r, r + \frac{R+\frac{\rho}{2}}{2\sqrt{1-\delta^2}}\right)$, which is the desired inequality. □

**Proposition B.5** *Let $(\varepsilon_j)$ be a sequence of positive numbers converging to 0, and let $E_j$ be such that $\sup_j F_{\varepsilon_j}(E_j) < +\infty$. Then, up to a subsequence, there exists $E$ such that $E_j \to E$, and for each $R > 0$ $E \cap Q(0,R)$ is a finite union of coordinate rectangles. If $|E| < +\infty$ then $E$ is a finite union of coordinate rectangles, and $F(E) \leq \liminf_j F_{\varepsilon_j}(E_j)$.*

**Proof** We consider first the case when each $E_j$ is connected. Proposition B.2 shows that if $\delta$ is sufficiently small, then $\#(K^j_\delta)$ is smaller than a constant $\widetilde{C}$, independent of $j$ and $\delta$. Let $(\delta_j)$ be a sequence converging to 0 chosen in such a way that $|I^j_{\delta_j}| \to 0$, and set

$$f(R) = \liminf_j \# \Big\{ k \in K^j_{\delta_j} : \gamma_j(I^{k,j}_{\delta_j}) \subseteq Q(0,R) \Big\}.$$

As $f(R)$ is a bounded, integer valued, increasing function, there exists $R_0$ such that

$$f(R_0) = \lim_{R\to+\infty} f(R).$$

Passing to a subsequence, from Remark B.3 we can assume that there exist $z_1, \dots z_N \in Q(0, R_0)$ such that $\lim_j \operatorname{dist}(\gamma_j(I_\delta^{k,j}), z_k) = 0$ for $k = 1, \dots, N$, while $\lim_j \gamma_j(I_\delta^{k,j}) = \infty$ for all other $k \in K_{\delta_j}^j$. Define

$$A = \{(x_1, x_2) \in \mathbf{R}^2 : \ x_1 = (z_k)_1 \text{ or } x_2 = (z_k)_2 \text{ for some } k = 1, \dots, N\}.$$

From the proof of Proposition B.4, for every $R > R_0$ and every $\rho > 0$, there exist $j_0$ such that

$$\partial E_j \cap Q(0, R) \subseteq A + B_\rho(0).$$

Moreover, every connected component of $Q(0, R) \setminus (A + B_\rho(0))$ is eventually contained in $E_j$ or in $\mathbf{R}^2 \setminus E_j$. Up to passing to a further subsequence we can suppose then that $E_j \to E$, where $E \cap Q(0, R)$ is the union of components of $Q(0, R) \setminus A$, hence a union of coordinate rectangles. Letting now $R \to +\infty$, we get that $E$ is union of components of $\mathbf{R}^2 \setminus A$.

If $|E| < +\infty$ then $E$ itself must be a union of bounded components of $\mathbf{R}^2 \setminus A$; i.e., a coordinate polyrectangle. In this case, we get from the construction above that $E$ has at most $N$ corners, one for each $z_k$. Hence, by Proposition B.2

$$F(E) \le c_W N \le \liminf_j (F_{\varepsilon_j}(E_j) + O(\sqrt{\delta_j}) = \liminf_j (F_{\varepsilon_j}(E_j) + O(\sqrt{\delta_j}).$$

In the case when $E_j$ are not connected, we see that the number of connected components is bounded by $4c_W \sup_j F_{\varepsilon_j}(E_j)$. Hence we can suppose that the number of components is $K$ independent of $j$. Denote by $E_j^1, \dots, E_j^K$ the connected components of $E_j$. Applying the reasoning above, we can suppose that $E_j^k \to E^k$, and for each $R > 0$ $E^k \cap Q(0, R)$ is a finite union of coordinate rectangles. We define then $E = \bigcup_{k=1}^K E^k$. Clearly $E \cap Q(0, R)$ is a finite union of coordinate rectangles. Finally, if $|E| < +\infty$, then it is a coordinate polyrectangle, and

$$\begin{aligned} F(E) &\le \sum_{k=1}^K F(E^k) \le \sum_{k=1}^K \liminf_j F_{\varepsilon_j}(E_j^k) \\ &\le \liminf_j \sum_{k=1}^K F_{\varepsilon_j}(E_j^k) = \liminf_j F_{\varepsilon_j}(E_j), \end{aligned}$$

as required. □

**Proposition B.6** *Let $E$ be a finite union of coordinate rectangles of $\mathbf{R}^2$; then there exists a family $(E_\varepsilon)$ of $C^2$ subsets of $\mathbf{R}^2$ such that $|E_\varepsilon \Delta E| \to 0$ and $F(E) \ge \limsup_{\varepsilon\to0^+} F_\varepsilon(E_\varepsilon)$.*

**Proof** Consider the solution $u$ of the Cauchy problem

$$\begin{cases} u' = \sqrt{W(u)(1-u^2)} \\ u(0) = \frac{1}{2}. \end{cases}$$

By using a cut-off argument, for all $T > 0$ a $C^1$ function $u_T : \mathbf{R} \to [0,1]$ can be defined such that

$$\begin{array}{ll} u_T(t) = u(x) & \text{if } |t| < T \\ u_T(t) = 0 & \text{if } t < -(T+1) \\ u_T(t) = 1 & \text{if } t > (T+1) \end{array}$$

and $|u_T'(t)| \leq \omega(T)$ if $T < |t| < T+1$, where $\omega(T) = o(1)$ as $T \to +\infty$. We then set

$$\gamma_\varepsilon(t) = \left(\int_0^t u_T\left(\frac{s}{\varepsilon}\right) ds, \int_0^t \sqrt{1 - u_T^2\left(\frac{s}{\varepsilon}\right)}\, ds\right).$$

$\gamma_\varepsilon$ is a curve parameterized by arc-length; moreover, $\dot{\gamma}_\varepsilon(t) = (1,0)$ if $t > (T+1)\varepsilon$ and $\dot{\gamma}_\varepsilon(t) = (0,1)$ if $t < -(T+1)\varepsilon$.

We get

$$\begin{aligned} &\int_{-\infty}^{+\infty} \left(\frac{1}{\varepsilon}\varphi(\dot{\gamma}_\varepsilon(t)) + \varepsilon|\ddot{\gamma}_\varepsilon(t)|^2\right) dt \\ &= \int_{-(T+1)}^{(T+1)} (W(u_T(t)) + |u_T'(t)|^2)\, dt \\ &\leq \max_{t \in [-(T+1), -T]} W(u_T(t)) + \max_{t \in [T, T+1]} W(u_T(t)) + 2\omega^2(T) \\ &\qquad + 2\int_{-T}^{T} \sqrt{\frac{W(u(t))}{1-u^2(t)}}\, u'(t)\, dt \\ &\leq 2\int_0^1 \sqrt{\frac{W(\tau)}{1-\tau^2}}\, d\tau + o(1) \end{aligned}$$

as $T \to +\infty$.

If $E$ is a finite union of coordinate rectangles, of vertices $x_i$, it is easy to construct the desired sequence $E_\varepsilon$ by choosing sets whose boundary is composed by arcs parameterized by $x_i \pm \gamma_\varepsilon$. □

Combining Propositions B.5 and B.6 we obtain a proof of Theorem B.1.

# APPENDIX C

## AN INTEGRAL REPRESENTATION RESULT

The ideas of Chapter 4 of deducing the form of the Γ-limit by a localization procedure and by the knowledge of its behaviour on a dense set of functions are at the core of the following integral representation result, which gives necessary and sufficient conditions for a functional to be represented as an integral on $SBV$ with Carathéodory integrands and given growth. This theorem may be useful in situations where it is difficult to apply directly the methods of Chapter 4 (see e.g. [BDV], [Co] or [CT]). We refer to [BCP] for its proof.

We define $SBV_p(\Omega)$ as the set of functions $u$ in $SBV(\Omega)$ such that $\nabla u \in L^p(\Omega;\mathbf{R}^n)$ and $\mathcal{H}^{n-1}(S(u)) < +\infty$.

**Theorem C.1** *Let* $F : SBV_p(\Omega) \times \mathcal{B}(\Omega) \to [0,+\infty)$ *be a functional satisfying the following conditions:*
(i) (locality on $\mathcal{A}(\Omega)$) *if* $u = v$ *a.e. on* $A \in \mathcal{A}(\Omega)$ *then* $F(u,A) = F(v,A)$*;*
(ii) (measure property) *for every* $u \in SBV_p(\Omega)$ *the set function* $B \mapsto F(u,B)$ *is a Borel measure;*
(iii) (lower semicontinuity) *for all* $A \in \mathcal{A}(\Omega)$ *the functional* $F(\cdot,A)$ *is lower semicontinuous on* $SBV_p(\Omega)$ *with respect to the* $\mathrm{L}^1(\Omega)$ *convergence;*
(iv) (growth condition of order $p$) *there exist* $a \in \mathrm{L}^1(\Omega)$ *such that*

$$0 \le F(u,B) \le \int_B (a(x) + |\nabla u|^p)\,dx + \mathcal{H}^{n-1}(S(u)\cap B) + \int_{S(u)\cap B} |u^+ - u^-|\,d\mathcal{H}^{n-1}$$

*for all* $u \in SBV_p(\Omega)$ *and* $B \in \mathcal{B}(\Omega)$*;*
(v) ("weak $\omega$ condition") *there exists a sequence* $(\omega_k)$ *of integrable moduli of continuity such that*

$$|F(u+s,A) - F(u,A)| \le \int_A \omega_k(x,|s|)\,dx$$

*for every* $k \in \mathbf{N}$, $A \in \mathcal{A}(\Omega)$, $s \in \mathbf{R}^m$, $u \in C^1$ *such that* $\|u\|_\infty \le k$, $\|u+s\|_\infty \le k$ *and* $\|Du\|_\infty \le k$*;*
(vi) (continuity of the jump energy) *there exists a modulus of continuity* $\omega$ *such that*

$$|F(u,S) - F(v,S)| \le \int_S \omega(|u^+ - v^+| + |u^- - v^-|)d\mathcal{H}^{n-1},$$

*for all* $u,v \in SBV_p(\Omega)$ *and* $S \subset S(u)\cap S(v)$ *(we choose the orientation* $\nu_v = \nu_u$ $\mathcal{H}^{n-1}$*-a.e. on* $S(u)\cap S(v)$*).*

*Then there exist Carathéodory functions* $f : \Omega \times \mathbf{R} \times \mathbf{R}^n \to [0,+\infty)$ *and* $\varphi : \Omega \times \mathbf{R} \times \mathbf{R} \times S^{n-1} \to [0,+\infty)$ *such that*

$$F(u,B) = \int_B f(x,u(x),\nabla u(x))\,dx + \int_{S(u)\cap B} \varphi(x,u^+,u^-,\nu_u)d\mathcal{H}^{n-1}$$

*for all* $u \in SBV_p(\Omega)$ *and* $B \in \mathcal{B}(\Omega)$.

**Remark C.2** (a) Sometimes condition (vi) may not be easy to verify. An alternative condition, which is suitable for many applications and easier to handle, is the following one:

(vi)′ for all $u \in SBV_p(\Omega)$ and $a, b \in \mathbf{R}$

$$|F(u,S(u)\cap A) - F(au+b, S(u)\cap A)| \le \omega(|a-1|+|b|)\int_{S(u)\cap A} (1+|u^+|+|u^-|)d\mathcal{H}^{n-1}$$

for all $A \in \mathcal{A}(\Omega)$.

(b) The function $f$ can be defined simply by derivation as

$$f(x_0,u_0,\xi_0) = \limsup_{\rho\to 0+} \frac{F(u_0 + \langle \xi_0, x - x_0\rangle, B_\rho(x_0))}{\mathcal{L}_n(B_\rho)}.$$

Such a simple description for $\varphi$, substituting somehow $\mathcal{L}_n$ by $\mathcal{H}^{n-1}$, is not possible, and in general false. However, a more complex derivation formula for $\varphi$ can be given as follows. Let $x \in \mathbf{R}^n$, $\rho > 0$, $\nu \in S^{n-1}$. We denote by $Q^\nu_\rho(x)$ an open cube centered in $x$ of side length $\rho$ and one face orthogonal to $\nu$. We will suppose that fixed $x$ and $\nu$ for each $\rho$ and $\sigma > 0$ the cube $Q^\nu_\sigma(x)$ is obtained from $Q^\nu_\rho(x)$ by an homothety of center $x$. Moreover, given $a$, $b \in \mathbf{R}$, we set

$$u^{\nu,x}_{a,b}(y) = \begin{cases} a & \text{if } \langle y-x,\nu\rangle > 0 \\ b & \text{if } \langle y-x,\nu\rangle \le 0. \end{cases}$$

Then $\varphi(x,a,b,\nu)$ is given by

$$\varphi(x,a,b,\nu) = \limsup_{\rho\to 0+} \frac{1}{\rho^{n-1}} \min\Big\{G(w,\overline{Q^\nu_\rho(x)}) : w \in SBV_p(\Omega),$$
$$\nabla u = 0 \text{ a.e.}, \; w = u^{\nu,x}_{a,b} \text{ on } \Omega\setminus Q^\nu_\rho(x)\Big\}$$

for all $x \in \Omega$, $a$, $b \in \mathbf{R}$, $\nu \in S^{n-1}$.

(c) The value of $\varphi$ on the set $\Omega \times \Delta \times S^{n-1}$, where $\Delta = \{(a,a) : \; a \in \mathbf{R}\}$ is the "diagonal" of $\mathbf{R}\times\mathbf{R}$, is never taken into account. Hence, the Carathéodory condition for $\varphi$ means that $\varphi(\cdot,a,b,\nu)$ is measurable for all $(a,b,\nu) \in \mathbf{R}\times\mathbf{R}\times S^{n-1}$ and $\varphi(x,\cdot,\cdot,\cdot)$ is continuous on $(\mathbf{R}\times\mathbf{R}\setminus\Delta)\times S^{n-1}$ for all $x \in \Omega$.

# APPENDIX D

## GAP PHENOMENON IN GSBV

When defining a functional $F$ on $GSBV(\Omega)$ from an expression which originally makes sense on $SBV(\Omega)$, the problem arises whether this extension is sensible, i.e. if the value given by $F$ on some function $u$ corresponds to some approximation of this function with functions $(u_j)$ of bounded variation, for which the functional has a precise meaning. In the words of the theory of relaxation, this can be expressed by the question: does the functional $F$ coincide with the lower semicontinuous envelope of its restriction on $SBV(\Omega)$? The answer is in general negative, and the functional must be "corrected" by an extra term. The "relaxed" functional $\overline{F}$ can be expressed in many cases as $F(u) + L(u)$, where the functional $L$ is not trivially 0, and it is given by an explicit formula. In this appendix we outline a special yet meaningful 1-dimensional model case, when it is possible to give a precise description of the relaxation. We fix an interval of $\mathbf{R}$, that we can take without loss of generality to be $I = (-1, 1)$, and we focus our attention on the behaviour near a point of $I$ where we may get a "degenerate" behaviour. Again, we can suppose that this point is 0.

We define the space of functions

$$GSBV_0(I) = GSBV(I) \cap SBV_{\rm loc}(I \setminus \{0\}),$$

and a convergence on $GSBV_0(I)$, by saying that a sequence $(u_j) \subset GSBV_0(I)$ converges to $u$ in $GSBV_0(I)$ if $u_j \rightharpoonup u$ weakly in $BV_{\rm loc}(I \setminus \{0\})$. Note that this convergence implies trivially a.e. convergence on $I$. The following theorem can be found in [Br3].

**Theorem D.1** *Let $f, g : \mathbf{R} \to [0, +\infty[$ be functions satisfying*

(a) *$f$ is convex;*

(b) *$g$ is lower semicontinuous and subadditive;*

(c) $\lim_{z\to\pm\infty} \frac{f(z)}{|z|} = \lim_{z\to 0} \frac{g(z)}{|z|} = +\infty$.

*Define $F$ on $SBV(I)$ by setting*

$$F(u) = \int_I f(u'(t))\,dt + \sum_{t\in S(u)\cap I} g(u(t_+) - u(t_-)),$$

*and $H$ by*

$$H(u) = \begin{cases} F(u) & \text{if } u \in SBV(I) \\ +\infty & \text{if } u \in GSBV_0(I) \setminus SBV(I). \end{cases}$$

*Then the lower semicontinuous envelope of $H$ in the topology of $GSBV_0(I)$ is given by*

$$\overline{H}(u) = F_0(u) + L(u)$$

*for all $u \in GSBV_0(I)$, where*

$$F_0(u) = \int_I f(u'(t))\,dt + \sum_{t \in S(u) \cap I \setminus \{0\}} g(u(t_+) - u(t_-)),$$

*and the* Lavrentiev term *$L$ is defined by*

$$L(u) = \liminf_{\varepsilon \to 0+} V(2\varepsilon, u(\varepsilon_+) - u(-\varepsilon_-)),$$

*with $V$ given by the* inf-convolution

$$V(x, s) = \min\left\{x\, f\left(\frac{s_1}{x}\right) + g(s_2) : \ s_1 + s_2 = s\right\}.$$

# NOTATION

*Sets, numbers*

$a \vee b$ ($a \wedge b$) the maximum (minimum) between $a$ and $b$

$A \Delta B$ the symmetric difference of $A$ and $B$

$A \subset\subset B$ means that the closure of $A$ is contained in the interior of $B$

$\mathcal{A}(\Omega)$ the family of open subsets of $\Omega$

$\mathcal{B}(\Omega)$ the family of Borel subsets of $\Omega$

$\mathcal{B}_c(\Omega)$ the family of Borel subsets of $\Omega$ with compact closure

$B_\rho(x)$ the open ball of centre $x$ and radius $\rho$

$c$ (if not otherwise stated) a strictly positive constant independent from the parameters of the problem, whose value may vary from line to line

$\Omega$ (if not otherwise stated) a bounded open subset of $\mathbf{R}^n$

$\langle \xi, \eta \rangle$ scalar product of $\xi$ and $\eta \in \mathbf{R}^N$

*Measures*

$|E|$ the Lebesgue measure of the set $E$

$\mathcal{H}^k$ the $k$-dimensional Hausdorff measure

$\mathcal{L}_n$ the Lebesgue measure in $\mathbf{R}^n$

$\mathcal{M}(\Omega; \mathbf{R}^N)$ the family of $\mathbf{R}^N$-valued measures on $\Omega$

$\mathcal{M}(\Omega)$ the family of scalar measures on $\Omega$

$\mathcal{M}_+(\Omega)$ the family of non-negative measures on $\Omega$

$\mu \llcorner E$ the restriction of the measure $\mu$ to $E$

$|\mu|$ the variation of the measure $\mu$

$\#$ the counting measure (number of elements of a set)

*Function spaces*

$BV(\Omega)$ the space of functions of bounded variation on $\Omega$

$C_c^k(\Omega; \mathbf{R}^N)$ the space of $\mathbf{R}^N$-valued functions with compact support in $\Omega$
$C_0^k(\Omega; \mathbf{R}^N)$ the space of $\mathbf{R}^N$-valued functions vanishing on $\partial\Omega$
($k$ omitted if 0; $\mathbf{R}^N$ omitted if $N = 1$)

$L^p(\Omega, \mu; \mathbf{R}^N)$ the space of $\mathbf{R}^N$-valued $p$-summable functions on $\Omega$ with respect to the measure $\mu$ ($\mu$ omitted if $\mathcal{L}_n$; $\mathbf{R}^N$ omitted if $N = 1$)

$W^{1,p}(\Omega)$ the space of Sobolev functions with $p$-summable derivatives on $\Omega$

*Functions*

$\chi_E$ the characteristic function of the set $E$

$Du$ the distributional derivative of $u$

$Lip(\phi)$ a Lipschitz constant for $\phi$

$\nabla u$ the approximate gradient of $u$

$\nu_u(x)$ the normal to $S(u)$ at $x$

$\rho$ a mollifier; $\rho_\gamma$ the scaled mollifier given by $\rho_\gamma(x) = \frac{1}{\gamma^n}\rho(\frac{x}{\gamma})$

$S(u)$ the complement of the set of Lebesgue points of $u$ (jump set)

$u^\pm(x)$ the approximate limits of $u$ at $x$

$u(t\pm)$ the right/left hand-side limits of $u$ at $t$

$\fint_B f\,dx$ the average of $f$ on $B$

# REFERENCES


[AB] E. Acerbi and A. Braides, Approximation of free-discontinuity problems by elliptic functionals via Γ-convergence, Preprint 1998.

[AM] G. Alberti and C. Mantegazza, A note on the theory of SBV functions, *Boll. Un. Mat. It. B* **11** (1997), 375-382.

[ABGe] R. Alicandro, A. Braides and M.S. Gelli, Free-discontinuity problems generated by singular perturbation, *Proc. Roy. Soc. Edinburgh*, to appear.

[ABS] R. Alicandro, A. Braides and J. Shah, Approximation of non-convex functionals in GBV, *Interfaces and Free Boundaries - modelling, analysis and computation*, to appear.

[AGe] R. Alicandro and M.S. Gelli, Free-discontinuity problems generated by singular perturbation: the $n$-dimensional case, Preprint 1998.

[AmB] M. Amar and A. Braides, A characterization of variational convergence for segmentation problems, *Discrete Continuous Dynamical Systems* **1** (1995), 347-369.

[A1] L. Ambrosio, A compactness theorem for a new class of functions of bounded variation, *Boll. Un. Mat. Ital.* **3-B** (1989), 857-881.

[A2] L. Ambrosio, Existence theory for a new class of variational problems, *Arch. Rational Mech. Anal.* **111** (1990), 291-322.

[A3] L. Ambrosio, On the lower semicontinuity of quasi-convex integrals defined in $SBV(\Omega;\mathbf{R}^k)$, *Nonlinear Anal.* **23** (1994), 405-425.

[A4] L. Ambrosio, A new proof of the $SBV$ compactness theorem, *Calc. Var.* **3** (1995), 127-137.

[A5] L. Ambrosio, *Corso introduttivo alla teoria geometrica della Misura ed alle Superfici Minime*, Scuola Normale Superiore Lecture Notes, Pisa, 1997.

[AB1] L. Ambrosio and A. Braides, Functionals defined on partitions of sets of finite perimeter, I: integral representation and Γ-convergence, *J. Math. Pures. Appl.* **69** (1990), 285-305.

[AB2] L. Ambrosio and A. Braides, Functionals defined on partitions of sets of finite perimeter, II: semicontinuity, relaxation and homogenization, *J. Math. Pures. Appl.* **69** (1990), 307-333.

[AB3] L. Ambrosio and A. Braides, Energies in SBV and variational models in fracture mechanics. *Homogenization and Applications to Material Sciences*, (D. Cioranescu, A. Damlamian, P. Donato eds.), GAKUTO, Gakkōtosho, Tokio, Japan, 1997, p. 1-22.

[ABG] L. Ambrosio, A. Braides and A. Garroni, Special functions with bounded variation and with weakly differentiable traces on the jump set, *Nonlinear Diff. Equations Appl.* **5** (1998), 219-243.

[ADM] L. Ambrosio and G. Dal Maso, A general chain rule for distributional derivatives, *Proc. Am. Math. Soc.* **108** (1990), 691-702.

[AFP] L. Ambrosio, N. Fusco and D. Pallara, Partial regularity of free-discontinuity sets II, *Ann. Sc. Norm. Super. Pisa, Cl. Sci., IV. Ser.* **24** (1997), 39-62.

[AFP1] L. Ambrosio, N. Fusco and D. Pallara, *Special Functions of Bounded Variation and Free Discontinuity Problems*, Oxford University Press, Oxford, to appear.

[AP] L. Ambrosio and D. Pallara, Partial regularity of free-discontinuity sets I, *Ann. Sc. Norm. Super. Pisa, Cl. Sci., IV. Ser.* **24** (1997), 1-38.

[AT1] L. Ambrosio and V. M. Tortorelli, Approximation of functionals depending on jumps by elliptic functionals via Γ-convergence, *Comm. Pure Appl. Math.* **43** (1990), 999-1036.

[AT2] L. Ambrosio and V. M. Tortorelli, On the approximation of free-discontinuity problems, *Boll. Un. Mat. Ital.* **6**-B (1992), 105-123.

[Ba] G.I. Barenblatt, The mathematical theory of equilibrium cracks in brittle fracture, *Adv. Appl. Mech.* **7** (1962), 55-129.

[BeC] G. Bellettini and A. Coscia, Discrete approximation of a free discontinuity problem, *Numer. Funct. Anal. Optim.* **15** (1994), 201-224.

[Bo] A. Bonnet, On the regularity of edges in image segmentation, *Ann. Inst. H. Poincaré Anal. Nonlin.* **13** (1996), 485-528.

[Br1] A. Braides, Homogenization of bulk and surface energies, *Boll. Un. Mat. Ital. B* **9** (1995), 375-398.

[Br2] A. Braides, Lower semicontinuity conditions for functionals on jumps and creases. *SIAM J. Math. Anal.* **26** (1995), 1184-1198.

[Br3] A. Braides, The Lavrentiev phenomenon for free discontinuity problems, *J. Funct. Anal.* **127** (1995), 1-20.

[BCP] A. Braides and V. Chiadò Piat, Integral representation results for functionals defined on $SBV(\Omega;\mathbf{R}^m)$, *J. Math. Pures Appl.* **75** (1996), 595-626.

[BC] A. Braides and A. Coscia, A singular perturbation approach to problems in fracture mechanics, *Math. Mod. Meth. Appl. Sci.* **3** (1993), 302-340.

[BDM] A. Braides and G. Dal Maso, Nonlocal approximation of the Mumford-Shah functional, *Calc. Var.* **5** (1997), 293-322.

[BDF] A. Braides and A. Defranceschi, *Homogenization of Multiple Integrals*, Oxford University Press, Oxford, 1998.

[BDV] A. Braides, A. Defranceschi and E. Vitali, Homogenization of free-discontinuity problems, *Arch. Rational Mech. Anal.* **135** (1996), 297-356.

[BG] A. Braides and A. Garroni, On the nonlocal approximation of free-discontinuity problems, *Comm. Partial. Diff. Equations*, to appear.

[BGe] A. Braides and M.S. Gelli, Limits of discrete systems with long-range interactions, Preprint 1998.

[BM] A. Braides and A. Malchiodi, Approximation of polyhedral energies, Preprint 1998.

[Bu] G. Buttazzo, *Semicontinuity, Relaxation and Integral Representation in the Calculus of Variations*, Longman, Harlow, 1989.

[CL] M. Carriero and A. Leaci, $S^k$-valued maps minimizing the $L^p$-norm of the gradient with free-discontinuities, *Ann. Scuola Norm. Sup. Pisa ser. IV* **18** (1991), 321-352.

[Ch] A. Chambolle, Image segmentation by variational methods: Mumford and Shah functional and the discrete approximations, *SIAM J. Appl. Math.* **55** (1995), 827-863.

[CTa] G. Congedo and I. Tamanini, On the existence of solutions to a problem in multidimensional segmentation, *Ann. Inst. H. Poincaré Anal. Non Linéaire* **2** (1991), 175-195.

[Co] G. Cortesani, Sequences of non-local functionals which approximate free-discontinuity problems, *Arch. Rational Mech. Anal.*, to appear.

[Co1] G. Cortesani, A finite element approximation of an image segmentation problem, *Math. Models Methods Appl. Sci.*, to appear.

[CT] G. Cortesani and R. Toader, A density result in *SBV* with respect to non-isotropic energies, *Nonlinear Anal.*, to appear.

[CT1] G. Cortesani and R. Toader, Non-local approximation of non-isotropic free-discontinuity problems, *SIAM J. Appl. Math.*, to appear.

[DM] G. Dal Maso, *An Introduction to $\Gamma$-convergence*, Birkhäuser, Boston, 1993.

[DMS] G. Dal Maso, J. M. Morel and S. Solimini, A variational method in image segmentation: existence and approximation results, *Acta Math.* **168** (1992), 89-151.

[DG] E. De Giorgi, Free Discontinuity Problems in Calculus of Variations, in: *Frontiers in pure and applied Mathematics, a collection of papers dedicated to J.L.Lions on the occasion of his 60th birthday* (R. Dautray ed.), North Holland, 1991.

[DGA] E. De Giorgi and L. Ambrosio, Un nuovo funzionale del calcolo delle variazioni, *Atti Accad. Naz. Lincei Rend. Cl. Sci. Fis. Mat. Natur.* **82** (1988), 199-210.

[DGC] E. De Giorgi, M. Carriero, and A. Leaci, Existence theorem for a minimum problem with free discontinuity set, *Arch. Rational Mech. Anal.* **108** (1989), 195-218.

[DGF] E. De Giorgi and T. Franzoni, Su un tipo di convergenza variazionale, *Atti Accad. Naz. Lincei Rend. Cl. Sci. Fis. Mat. Natur.* **58** (1975), 842-850.

[Di] F. Dibos, Uniform rectifiability of image segmentations obtained by a variational method, *J. Math. Pures Appl.*, to appear.

[EG] L. C. Evans and R. F. Gariepy, *Measure theory and fine properties of functions*, CRC Press, Boca Raton 1992.

[Fe] H. Federer, *Geometric Measure Theory.* Springer-Verlag, New York, 1969.

[FF] I. Fonseca and G. Francfort, A model for the interaction between fracture and damage, *Calc. Var.* **3** (1995), 407-446.

[Gi] E. Giusti, *Minimal Surfaces and Functions of Bounded Variation*, Birkhäuser, Basel 1983.

[Go] M. Gobbino, Finite difference approximation of the Mumford-Shah functional, *Comm. Pure Appl. Math.* **51** (1998), 197-228.

[GM] M. Gobbino and M.G. Mora, Finite difference approximation of free discontinuity problems, Preprint 1998.

[MM] L. Modica and S. Mortola, Un esempio di Γ-convergenza, *Boll. Un. Mat. It. B* **14** (1977), 285-299.

[MoS] J. M. Morel and S. Solimini, *Variational Models in Image Segmentation*, Birkhäuser, Boston, 1995.

[Mu] D. Mumford, The statistical description of visual signals, in *ICIAM 95: Proceedings of the third international congress on industrial and applied mathematics, held in Hamburg, Germany, July 3-7, 1995* (K. Kirchgaessner, ed.), Akademie Verlag, Berlin, *Math. Res.* **87** (1996), 233-256.

[MS] D. Mumford and J. Shah, Optimal approximation by piecewise smooth functions and associated variational problems, *Comm. Pure Appl. Math.* **17** (1989), 577-685.

[Sh] J. Shah, A common framework for curve evolution, segmentation and anisotropic diffusion, *IEEE Conference on Computer Vision and Pattern Recognition*, June, 1996.

[St] H. Stark, *Image Recovery: Theory and Application*, Academic Press, Orlando, 1987.

[Vi] E.G. Virga, Drops of nematic liquid crystals, *Arch. Rational Mech Anal.* **107** (1989), 371-390.

# INDEX

Printing: Weihert-Druck GmbH, Darmstadt
Binding: Buchbinderei Schäffer, Grünstadt

Vol. 1596: L. Heindorf, L. B. Shapiro, Nearly Projective Boolean Algebras. X, 202 pages. 1994.

Vol. 1597: B. Herzog, Kodaira-Spencer Maps in Local Algebra. XVII, 176 pages. 1994.

Vol. 1598: J. Berndt, F. Tricerri, L. Vanhecke, Generalized Heisenberg Groups and Damek-Ricci Harmonic Spaces. VIII, 125 pages. 1995.

Vol. 1599: K. Johannson, Topology and Combinatorics of 3-Manifolds. XVIII, 446 pages. 1995.

Vol. 1600: W. Narkiewicz, Polynomial Mappings. VII, 130 pages. 1995.

Vol. 1601: A. Pott, Finite Geometry and Character Theory. VII, 181 pages. 1995.

Vol. 1602: J. Winkelmann, The Classification of Three-dimensional Homogeneous Complex Manifolds. XI, 230 pages. 1995.

Vol. 1603: V. Ene, Real Functions – Current Topics. XIII, 310 pages. 1995.

Vol. 1604: A. Huber, Mixed Motives and their Realization in Derived Categories. XV, 207 pages. 1995.

Vol. 1605: L. B. Wahlbin, Superconvergence in Galerkin Finite Element Methods. XI, 166 pages. 1995.

Vol. 1606: P.-D. Liu, M. Qian, Smooth Ergodic Theory of Random Dynamical Systems. XI, 221 pages. 1995.

Vol. 1607: G. Schwarz, Hodge Decomposition – A Method for Solving Boundary Value Problems. VII, 155 pages. 1995.

Vol. 1608: P. Biane, R. Durrett, Lectures on Probability Theory. Editor: P. Bernard. VII, 210 pages. 1995.

Vol. 1609: L. Arnold, C. Jones, K. Mischaikow, G. Raugel, Dynamical Systems. Montecatini Terme, 1994. Editor: R. Johnson. VIII, 329 pages. 1995.

Vol. 1610: A. S. Üstünel, An Introduction to Analysis on Wiener Space. X, 95 pages. 1995.

Vol. 1611: N. Knarr, Translation Planes. VI, 112 pages. 1995.

Vol. 1612: W. Kühnel, Tight Polyhedral Submanifolds and Tight Triangulations. VII, 122 pages. 1995.

Vol. 1613: J. Azéma, M. Emery, P. A. Meyer, M. Yor (Eds.), Séminaire de Probabilités XXIX. VI, 326 pages. 1995.

Vol. 1614: A. Koshelev, Regularity Problem for Quasilinear Elliptic and Parabolic Systems. XXI, 255 pages. 1995.

Vol. 1615: D. B. Massey, Le Cycles and Hypersurface Singularities. XI, 131 pages. 1995.

Vol. 1616: I. Moerdijk, Classifying Spaces and Classifying Topoi. VII, 94 pages. 1995.

Vol. 1617: V. Yurinsky, Sums and Gaussian Vectors. XI, 305 pages. 1995.

Vol. 1618: G. Pisier, Similarity Problems and Completely Bounded Maps. VII, 156 pages. 1996.

Vol. 1619: E. Landvogt, A Compactification of the Bruhat-Tits Building. VII, 152 pages. 1996.

Vol. 1620: R. Donagi, B. Dubrovin, E. Frenkel, E. Previato, Integrable Systems and Quantum Groups. Montecatini Terme, 1993. Editors:M. Francaviglia, S. Greco. VIII, 488 pages. 1996.

Vol. 1621: H. Bass, M. V. Otero-Espinar, D. N. Rockmore, C. P. L. Tresser, Cyclic Renormalization and Auto-morphism Groups of Rooted Trees. XXI, 136 pages. 1996.

Vol. 1622: E. D. Farjoun, Cellular Spaces, Null Spaces and Homotopy Localization. XIV, 199 pages. 1996.

Vol. 1623: H.P. Yap, Total Colourings of Graphs. VIII, 131 pages. 1996.

Vol. 1624: V. Brınzanescu, Holomorphic Vector Bundles over Compact Complex Surfaces. X, 170 pages. 1996.

Vol.1625: S. Lang, Topics in Cohomology of Groups. VII, 226 pages. 1996.

Vol. 1626: J. Azéma, M. Emery, M. Yor (Eds.), Séminaire de Probabilités XXX. VIII, 382 pages. 1996.

Vol. 1627: C. Graham, Th. G. Kurtz, S. Méléard, Ph. E. Protter, M. Pulvirenti, D. Talay, Probabilistic Models for Nonlinear Partial Differential Equations. Montecatini Terme, 1995. Editors: D. Talay, L. Tubaro. X, 301 pages. 1996.

Vol. 1628: P.-H. Zieschang, An Algebraic Approach to Association Schemes. XII, 189 pages. 1996.

Vol. 1629: J. D. Moore, Lectures on Seiberg-Witten Invariants. VII, 105 pages. 1996.

Vol. 1630: D. Neuenschwander, Probabilities on the Heisenberg Group: Limit Theorems and Brownian Motion. VIII, 139 pages. 1996.

Vol. 1631: K. Nishioka, Mahler Functions and Transcendence.VIII, 185 pages.1996.

Vol. 1632: A. Kushkuley, Z. Balanov, Geometric Methods in Degree Theory for Equivariant Maps. VII, 136 pages. 1996.

Vol.1633: H. Aikawa, M. Essén, Potential Theory – Selected Topics. IX, 200 pages.1996.

Vol. 1634: J. Xu, Flat Covers of Modules. IX, 161 pages. 1996.

Vol. 1635: E. Hebey, Sobolev Spaces on Riemannian Manifolds. X, 116 pages. 1996.

Vol. 1636: M. A. Marshall, Spaces of Orderings and Abstract Real Spectra. VI, 190 pages. 1996.

Vol. 1637: B. Hunt, The Geometry of some special Arithmetic Quotients. XIII, 332 pages. 1996.

Vol. 1638: P. Vanhaecke, Integrable Systems in the realm of Algebraic Geometry. VIII, 218 pages. 1996.

Vol. 1639: K. Dekimpe, Almost-Bieberbach Groups: Affine and Polynomial Structures. X, 259 pages. 1996.

Vol. 1640: G. Boillat, C. M. Dafermos, P. D. Lax, T. P. Liu, Recent Mathematical Methods in Nonlinear Wave Propagation. Montecatini Terme, 1994. Editor: T. Ruggeri. VII, 142 pages. 1996.

Vol. 1641: P. Abramenko, Twin Buildings and Applications to S-Arithmetic Groups. IX, 123 pages. 1996.

Vol. 1642: M. Puschnigg, Asymptotic Cyclic Cohomology. XXII, 138 pages. 1996.

Vol. 1643: J. Richter-Gebert, Realization Spaces of Polytopes. XI, 187 pages. 1996.

Vol. 1644: A. Adler, S. Ramanan, Moduli of Abelian Varieties. VI, 196 pages. 1996.

Vol. 1645: H. W. Broer, G. B. Huitema, M. B. Sevryuk, Quasi-Periodic Motions in Families of Dynamical Systems. XI, 195 pages. 1996.

Vol. 1646: J.-P. Demailly, T. Peternell, G. Tian, A. N. Tyurin, Transcendental Methods in Algebraic Geometry. Cetraro, 1994. Editors: F. Catanese, C. Ciliberto. VII, 257 pages. 1996.

Vol. 1647: D. Dias, P. Le Barz, Configuration Spaces over Hilbert Schemes and Applications. VII. 143 pages. 1996.

Vol. 1648: R. Dobrushin, P. Groeneboom, M. Ledoux, Lectures on Probability Theory and Statistics. Editor: P. Bernard. VIII, 300 pages. 1996.

Vol. 1649: S. Kumar, G. Laumon, U. Stuhler, Vector Bundles on Curves – New Directions. Cetraro, 1995. Editor: M. S. Narasimhan. VII, 193 pages. 1997.

Vol. 1650: J. Wildeshaus, Realizations of Polylogarithms. XI, 343 pages. 1997.

Vol. 1651: M. Drmota, R. F. Tichy, Sequences, Discrepancies and Applications. XIII, 503 pages. 1997.

Vol. 1652: S. Todorcevic, Topics in Topology. VIII, 153 pages. 1997.

Vol. 1653: R. Benedetti, C. Petronio, Branched Standard Spines of 3-manifolds. VIII, 132 pages. 1997.

Vol. 1654: R. W. Ghrist, P. J. Holmes, M. C. Sullivan, Knots and Links in Three-Dimensional Flows. X, 208 pages. 1997.

Vol. 1655: J. Azéma, M. Emery, M. Yor (Eds.), Séminaire de Probabilités XXXI. VIII, 329 pages. 1997.

Vol. 1656: B. Biais, T. Björk, J. Cvitanic, N. El Karoui, E. Jouini, J. C. Rochet, Financial Mathematics. Bressanone, 1996. Editor: W. J. Runggaldier. VII, 316 pages. 1997.

Vol. 1657: H. Reimann, The semi-simple zeta function of quaternionic Shimura varieties. IX, 143 pages. 1997.

Vol. 1658: A. Pumarino, J. A. Rodrıguez, Coexistence and Persistence of Strange Attractors. VIII, 195 pages. 1997.

Vol. 1659: V, Kozlov, V. Maz'ya, Theory of a Higher-Order Sturm-Liouville Equation. XI, 140 pages. 1997.

Vol. 1660: M. Bardi, M. G. Crandall, L. C. Evans, H. M. Soner, P. E. Souganidis, Viscosity Solutions and Applications. Montecatini Terme, 1995. Editors: I. Capuzzo Dolcetta, P. L. Lions. IX, 259 pages. 1997.

Vol. 1661: A. Tralle, J. Oprea, Symplectic Manifolds with no Kähler Structure. VIII, 207 pages. 1997.

Vol. 1662: J. W. Rutter, Spaces of Homotopy Self-Equivalences – A Survey. IX, 170 pages. 1997.

Vol. 1663: Y. E. Karpeshina; Perturbation Theory for the Schrödinger Operator with a Periodic Potential. VII, 352 pages. 1997.

Vol. 1664: M. Väth, Ideal Spaces. V, 146 pages. 1997.

Vol. 1665: E. Giné, G. R. Grimmett, L. Saloff-Coste, Lectures on Probability Theory and Statistics 1996. Editor: P. Bernard. X, 424 pages, 1997.

Vol. 1666: M. van der Put, M. F. Singer, Galois Theory of Difference Equations. VII, 179 pages. 1997.

Vol. 1667: J. M. F. Castillo, M. González, Three-space Problems in Banach Space Theory. XII, 267 pages. 1997.

Vol. 1668: D. B. Dix, Large-Time Behavior of Solutions of Linear Dispersive Equations. XIV, 203 pages. 1997.

Vol. 1669: U. Kaiser, Link Theory in Manifolds. XIV, 167 pages. 1997.

Vol. 1670: J. W. Neuberger, Sobolev Gradients and Differential Equations. VIII, 150 pages. 1997.

Vol. 1671: S. Bouc, Green Functors and $G$-sets. VII, 342 pages. 1997.

Vol. 1672: S. Mandal, Projective Modules and Complete Intersections. VIII, 114 pages. 1997.

Vol. 1673: F. D. Grosshans, Algebraic Homogeneous Spaces and Invariant Theory. VI, 148 pages. 1997.

Vol. 1674: G. Klaas, C. R. Leedham-Green, W. Plesken, Linear Pro-$p$-Groups of Finite Width. VIII, 115 pages. 1997.

Vol. 1675: J. E. Yukich, Probability Theory of Classical Euclidean Optimization Problems. X, 152 pages. 1998.

Vol. 1676: P. Cembranos, J. Mendoza, Banach Spaces of Vector-Valued Functions. VIII, 118 pages. 1997.

Vol. 1677: N. Proskurin, Cubic Metaplectic Forms and Theta Functions. VIII, 196 pages. 1998.

Vol. 1678: O. Krupková, The Geometry of Ordinary Variational Equations. X, 251 pages. 1997.

Vol. 1679: K.-G. Grosse-Erdmann, The Blocking Technique. Weighted Mean Operators and Hardy's Inequality. IX, 114 pages. 1998.

Vol. 1680: K.-Z. Li, F. Oort, Moduli of Supersingular Abelian Varieties. V, 116 pages. 1998.

Vol. 1681: G. J. Wirsching, The Dynamical System Generated by the 3n+1 Function. VII, 158 pages. 1998.

Vol. 1682: H.-D. Alber, Materials with Memory. X, 166 pages. 1998.

Vol. 1683: A. Pomp, The Boundary-Domain Integral Method for Elliptic Systems. XVI, 163 pages. 1998.

Vol. 1684: C. A. Berenstein, P. F. Ebenfelt, S. G. Gindikin, S. Helgason, A. E. Tumanov, Integral Geometry, Radon Transforms and Complex Analysis. Firenze, 1996. Editors: E. Casadio Tarabusi, M. A. Picardello, G. Zampieri. VII, 160 pages. 1998.

Vol. 1685: S. König, A. Zimmermann, Derived Equivalences for Group Rings. X, 146 pages. 1998.

Vol. 1686: J. Azéma, M. Émery, M. Ledoux, M. Yor (Eds.), Séminaire de Probabilités XXXII. VI, 440 pages. 1998.

Vol. 1687: F. Bornemann, Homogenization in Time of Singularly Perturbed Mechanical Systems. XII, 156 pages. 1998.

Vol. 1688: S. Assing, W. Schmidt, Continuous Strong Markov Processes in Dimension One. XII, 137 page. 1998.

Vol. 1689: W. Fulton, P. Pragacz, Schubert Varieties and Degeneracy Loci. XI, 148 pages. 1998.

Vol. 1690: M. T. Barlow, D. Nualart, Lectures on Probability Theory and Statistics. Editor: P. Bernard. VIII, 237 pages. 1998.

Vol. 1691: R. Bezrukavnikov, M. Finkelberg, V. Schechtman, Factorizable Sheaves and Quantum Groups. X, 282 pages. 1998.

Vol. 1692: T. M. W. Eyre, Quantum Stochastic Calculus and Representations of Lie Superalgebras. IX, 138 pages. 1998.

Vol. 1694: A. Braides, Approximation of Free-Discontinuity Problems. XI, 149 pages. 1998.

# General Remarks

Lecture Notes are printed by photo-offset from the master-copy delivered in camera-ready form by the authors. For this purpose Springer-Verlag provides technical instructions for the preparation of manuscripts.

Careful preparation of manuscripts will help keep production time short and ensure a satisfactory appearance of the finished book. The actual production of a Lecture Notes volume normally takes approximately 8 weeks.

Authors receive 50 free copies of their book. No royalty is paid on Lecture Notes volumes.

Authors are entitled to purchase further copies of their book and other Springer mathematics books for their personal use, at a discount of 33,3 % directly from Springer-Verlag.

Commitment to publish is made by letter of intent rather than by signing a formal contract. Springer-Verlag secures the copyright for each volume.

Addresses:

Professor A. Dold
Mathematisches Institut
Universität Heidelberg
Im Neuenheimer Feld 288
D-69120 Heidelberg, Germany

Professor F. Takens
Mathematisch Instituut
Rijksuniversiteit Groningen
Postbus 800
NL-9700 AV Groningen
The Netherlands

Professor Bernard Teissier
École Normale Supérieure
45, rue d'Ulm
F-7500 Paris, France

Springer-Verlag, Mathematics Editorial
Tiergartenstr. 17
D-69121 Heidelberg, Germany
Tel.: *49 (6221) 487-410